Environmental Data Exchange Network for Inland Water

Environmental Data Exchange Network for Inland Water

Edited by:

Palle Haastrup
ISPRA
Institute for Environment and Sustainability
TP 290 Ispra (VA) 21020 Italy

Jørgen Würtz
ISPRA
Institute for Environment and Sustainability
TP 290 Ispra (VA) 21020 Italy

ELSEVIER

Amsterdam • Boston • Heidelberg • London • New York • Oxford
Paris • San Diego • San Francisco • Singapore • Sydney • Tokyo

Elsevier
Radarweg 29, PO Box 211, 1000 AE Amsterdam, The Netherlands
The Boulevard, Langford Lane, Kidlington, Oxford OX5 1GB, UK

First edition 2007

Library of Congress Cataloging-in-Publication Data
A catalog record for this book is available from the Library of Congress

British Library Cataloguing in Publication Data
A catalogue record for this book is available from the British Library

ISBN-13: 978-0-444-52973-2
ISBN-10: 0-444-52973-X

For information on all Elsevier publications
visit our website at books.elsevier.com

Printed and bound in The Netherlands

07 08 09 10 11 10 9 8 7 6 5 4 3 2 1

CONTENTS

Foreword vii
M. Grasserbauer, JRC Ispra
Preface ix
G. Weets, DG-INFSO
Acknowledgments xi
List of Contributors xiii

PART I Environmental Data Exchange 1

1 Environmental Data Exchange for Inland Water 3
P. Haastrup and J. Würtz

2 An Example of a Data Exchange Network: The French System for Water Data 17
D. Preux

3 The Evolution of Thematic Reporting into a Shared Information System: An Example on Water 29
S. Jensen

4 Agent-Based Concepts for Environmental Data 39
J. Fowler and G. Pitts

5 Review of Models and Technologies for Database Integration 51
S. Poslad, M. Stjernholm, L. Zuo, and X. Huang

6 Security Issues for Distributed Software Agents 69
A. Papageorgiou

PART II The EDEN-IW Project 81

7 Introduction and Overview of Results 83
P. Haastrup

8 Multi-agent System Technology in Distributed Database Systems 97
S. Poslad, L. Zuo, and X. Huang

the work by exploiting potential arenas for data sharing, which is an important topic for discussion among Member States and the Commission in relation to the reporting obligations.

Manfred Grasserbauer
Director
Institute for Environment and Sustainability
Joint Research Centre, Ispra
European Commission
September 2005

PREFACE

Understanding and protecting our environment is essential for the quality of life of future generations; it is also a key component of sustainable development and is becoming therefore part of other policies such as health, innovation and competitiveness.

Access to a wide range of high-quality information is essential to our ability to understand and safeguard the environment. Paradoxically, many environmental theories, assumptions and models are currently based on very limited amount of data. This is mainly due to the fact that there is virtually no exchange of data between boundaries – be they national, regional or even institutional. There are two major reasons for this: the first is the wide variety of data policies or the lack of it; the second is the heterogeneity of the data format and the lack of documentation.

The Environmental Data Exchange Network for Inland Water (EDEN-IW) project, which was completed at the end of 2004, addressed successfully the second issue.

The project was financed by the user-friendly information society (IST) programme under the Fifth European Community Framework Programme for Research, Technological Development and Demonstration activities. The strategic objective of this programme was to realise the benefits of the IST for Europe both by accelerating its emergence and by ensuring that the needs of citizens and enterprises were met.

The use of ICT (Information and Communication Technologies) in the field of environment is to provide decision makers with easy access to strategic information and models in support of their decisions in managing the environment. This ICT sector, known as eco-informatics, is growing rapidly and is becoming an integral part of the European eco-industry, whose turnover is estimated to be above 100 billion euros annually, with a growth rate of around 8%.

For the public sector (regional, national, European), ICT contributes to the improvement of a number of environmental policies from definition to monitoring. It helps to achieve world-class environment administrations in Europe by enhancing their efficiency, openness and accountability.

The EDEN-IW project addressed the key issues at the center of the strategy. Its objectives emphasised the environmental management system focused on generic scalable information systems covering water and pollution.

Many heterogeneous data are being collected by local, regional and national actors to monitor the quality of our fresh water, rivers, reservoirs and lakes. However, reporting quality at European level of the so-called "Inland Waters" requires integrated access to numerous databases around the EU. EDEN-IW provides such connectivity through what may be described as a virtual data warehouse, based on software agents which advertise, broker and retrieve data, thus allowing rapid and comprehensive answers to data queries. The EDEN-IW data exchange prototype is a direct RTD contribution to the implementation of, for instance, the EU's 6th Environmental Action Plan, the Water framework Directive from 2000, the Nitrate Directive from 1991, and the Drinking Water Directive from 1980.

The EDEN-IW prototype is generic and scalable, and involves distributed computing, advanced model simulation and forecasting, data-mining, advanced visualisation techniques and decision-support systems.

The results of the EDEN-IW project accelerate the emergence of the advanced information technology and expand its circle of users to a wide range of individuals and organisations.

A key driving force for the project was to meet the need and expectation of high-quality affordable services related to environmental data, initially on the Inland Water field, but with perspectives and visions that include more general information related to the environment.

An important focus of the IST programme was to improve interaction with applications and services, especially multi-lingual adaptable systems, within the environment fields – EDEN-IW is a good example.

Thus, it is with considerable pleasure that I write the preface to this book with its extended results of the successful Environmental Data Exchange Network for Inland Water project.

Guy Weets
Deputy Head of Unit
ICT for the environment
DG Information Society
European Commission

ACKNOWLEDGMENTS

This book is based on the results of a substantial research effort over a period of nearly five years, with significant financial contributions from the European Commission and various partners.

The research project followed on to previous research efforts both in Europe and US, and the current project would not have happened without significant help from many people.

The chapters in the book have been written by the following authors:

B. Bargmeyer (LBNL*); A.R. Cassandra (setenv); S. Chance (US-DOD); A. de Jager (JRC); B. Felluga (CNR); J. Fowler (setenv); M. Grasserbauer (JRC); P. Haastrup (JRC); X. Huang (QMUL); S. Jensen (EEA); R. Lozado (setenv); A. Papageorgiou (ED); G. Pitts (GEPA); P. Plini (CNR); S. Poslad (QMUL); D. Preux (IOW); F-X. Prunayre (IOW); E. Rimaviciute (JRC); O. Sortkjær (NERI); M. Stjernholm (NERI); G. Weets (DG-INFSO); J. Würtz (JRC); L. Zuo (QMUL).

Among the support and active contributions, we would like to acknowledge the help and support of the following colleagues, family and friends:

I. Allegrini (CNR); C. Baffioni (Italian Min. of Env.); J. Bigham (QMUL); W. Bock (DG-INFSO); A. Boschet (ETC/Water); M. Brdys (Tech. Univ. Gdansk); N. Brohan (DG-INFSO); F. Burbello (Doc-A-Ware); P. Collotte (DGINFSO); M. Cullen (US-EPA); G. Cunningham (UNEP); S. Dalal (Telcordia); V. De Santis (CNR); R. Denzer (Saarland State Univ.); A. Dimopoulos (ED); M. Dowell (JRC); S. Eisenreich (JRC); L. Fitzwater (US-EPA); C. Foutsitzis (ED); T. Gauthier (IOW); D. Gritsis (ED); A. Gunesh (QMUL); B. Haastrup, S. Haastrup, I. Heiber (DG-INFSO); T. Kavalieros (ED); T. Lack (ETC/Water); Y. Lazopoulos (ED); S. Lucke (CNR); G. Makris (ED); F. Marcos (Finmeccanica); G. Mark (US-DOD); J. Martin (EEA); J-M. Martin (JRC); P. Meozzi (EEA); T. Moth Iversen (NERI); M. Mulvenna (Univ. Ulster); P. Murphy (DG-ENV); R. Murra (US-DOE); C. Neophytou (ED); M. Nodine (MCC); B. Norup(†) (NERI); C. Pipitsoulis (DG-INFSO); Z. Protogeros (ED); P. Quevauviller (DG-ENV); K. Rasmussen (JRC), M. Rusinkiewicz (Telcordia); H. Saarenmaa (EEA); I. Santouridis (ED); W. Schrimpf (JRC); S. Scott (US-DOE); S. Simon (US-DOE); M Slater (US-DOD); W. Sonntag (EPA); P. Soprano (Italian Min. of Env.); L. Sortkjær (NERI); L. Spencer

*See the list of acronyms at the end of the book.

(US-EPA); P. Theodossopoulos (ED); R. Tolksdorf (Frie Univ. Berlin); A. Van Der Beken (Vrie Univ. Brussels); M. Viljanen (DG-INFSO); D. Wilkinson (JRC).

In addition, support was received from various database owners, among them are EEA (ETC Water), Environment Agency for England and Wales, IOW, NERI.

Palle Haastrup
Jørgen Würtz

LIST OF CONTRIBUTORS

Bruce Bargmeyer, Lawrence Berkeley National Laboratory, US

Anthony R. Cassandra, Setenv, US

Sam Chance, US-DOD, US

Alfred de Jager, JRC Ispra

Bruno Felluga, Consiglio Nazionale delle Ricerce, Italy

Jerry Fowler, Setenv, US

Manfred Grasserbauer, JRC Ispra

Palle Haastrup, JRC Ispra

Xuan Huang, Queen Mary University of London, UK

Stefan Jensen, European Environment Agency, Denmark

Ryan J. Lozado, Setenv, US

Athanassios Papageorgiou, European Dynamics, Greece and JRC Ispra

Greg Pitts, GEPA, US

Paolo Plini, Consiglio Nazionale delle Ricerce, Italy

Stefan Poslad, Queen Mary University of London, UK

Dominique Preux, International Office for Water, France

François-Xavier Prunayre, International Office for Water, France

Erika Rimaviciute, JRC Ispra

Ole Sortkjær, National Environmental Research Institute, Denmark

Michael Stjernholm, National Environmental, Research Institute, Denmark

Guy Weets, DG-INFSO

Jørgen Würtz, JRC Ispra

Landong Zuo, Queen Mary University of London, UK

PART I

ENVIRONMENTAL DATA EXCHANGE

Chapter 1

ENVIRONMENTAL DATA EXCHANGE FOR INLAND WATER

P. Haastrup and J. Würtz

1.1 INTRODUCTION

In a world where computing power, storage capacity and network size all tend to grow exponentially, exchange of data is growing at a similar rate. From a purely information-technology point of view, any data packet sent over the network is "exchange of data", and is thus linked to the overall network traffic.

Searching the Web for relevant information has become commonplace; indeed, it is at times difficult to imagine how cumbersome the search for information was before the World Wide Web.

It is amazing to see the ease with which it is now possible to search for and find, for instance, telephone numbers, addresses, flight information and scientific publications (just to mention a few). However, for other types of data, the search is more difficult and has yet to realise the full potential of the full data exchange.

On the philosophical level, one could speculate that the exchange of document information through the World Wide Web has been so successful due to the fact that the HyperText Markup language (HTML) was invented at the same time as the first browsers became operational and that the protocol and the language were so simple and easy to use.

For data held in databases, this was not the case, with different standards and proprietary software already in place on stand-alone computers, making data exchange an inherently difficult problem.

These general problems on data exchange are also relevant for data exchange in the environmental field, and in the field of Inland Water.

For the environmental topic of surface water, one additional physical characteristic is fundamental: water runs downhill (!) Environmental data for one water body upstream of another is therefore always relevant and allows the inference of some scientific and technical information about the water body downstream.

With this in mind, the Environmental Data Exchange Network for Inland Water (EDEN-IW) project (described in detail in Part II) was started, aiming to facilitate and increase data exchange.

The perspectives and the visions of the needs are that a citizen or a user, including policy makers, needs to be able to get rapid answers to environmental questions on inland waters, without any knowledge about what data sources might be available or how to connect to them. The questions may range from global to continental, from national to local, following political boundaries or catchment areas as required.

In principle, a large number of existing Inland Water databases in the world could be linked together, providing citizens with access to a worldwide inland water database, together with tools to assess the information and regulate the inland water environment in a more efficient and effective way than in the current situation. The present status is illustrated in Figure 1.1.

A key driving force for the project was to meet the needs and expectations of high-quality affordable services related to environmental data, initially on the Inland

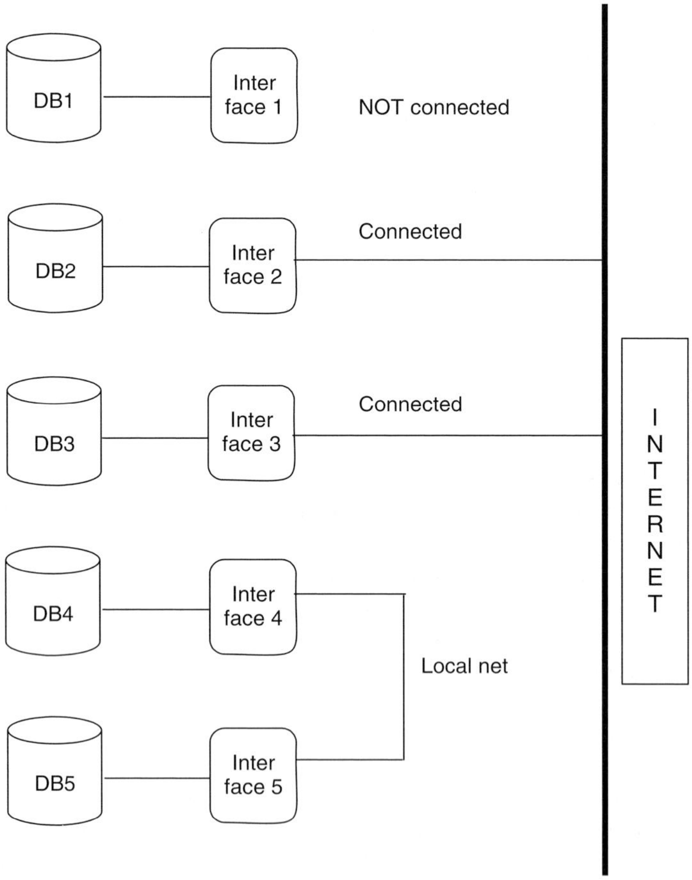

Fig. 1.1. Status of environmental databases.

Water field, but with perspectives and visions that include more general information related to the environment.

As illustrated in Figure 1.1, the current situation is a mix of all possible combinations, with some databases disconnected from the network, through situation with remote accessibility on the corporate or local area network, all the way to databases fully connected to the Internet. However, in the situation, where the databases are connected to the Internet, they typically have different interfaces and use different protocols.

The aim was to change this situation, to enable individuals and organisations to innovate and be more effective and efficient in their work related to environmental management, thereby providing the basis for sustainable growth while also improving the quality of life.

The goals of the project – with an integrative approach as a fundamental point of departure – are well adapted to the requirements for policy issues in areas where the information society technologies, by nature, require close co-ordination with various research programmes and where the deployment plays a critical role.

The challenges were and are significant. Previous experience, relevant technologies and technological advancements had to be examined, leading to research and development of specific information technology solutions.

The background is examined in the next sections, while previous experiences in environmental data exchange are examined in Chapter 2 with an example of a French water system, and the shared information of the EEA (Chapter 3).

1.2 GENERAL BACKGROUND

A significant number of public institutions and organisations gather, analyse, and disseminate environmental data from many sources. Traditionally, each organisation has developed its own databases with different database softwares and on different platforms: Each database addresses different aspects of the data. Both from a practical and from a theoretical point of view, it is important to discover inconsistencies in the data and minimise redundancies. Because the databases differ by their structure, query language, location, and by the way they represent data, it has been virtually impossible, even for a skilled database professional, to retrieve common information from these databases simultaneously. This isolation of environmental information limits its usefulness and increases the risk of inconsistency.

Many organisations gather environmental multilingual data from many sources and make them available to public and private users – though lack of consistent data descriptions has made it very difficult to share data and use them as a knowledge source in decision-making. The difficulty of organising and managing what we know about the data has led to the development of distributed systems that address isolated functions. However, the physical separation of these systems hinders or prevents access by secondary users and frustrates attempts to draw data together to form a more comprehensive understanding of environmental conditions and actions.

Previous research (US-EPA, 2005) has demonstrated the potential of the elements using distributed, heterogeneous information management by providing a conceptual demonstration of how the use of metadata registries, such as the Environmental Data Registry (EDR) of the US-EPA, can be combined with thesaurus systems, such as the Terminology Reference System (TRS) (also of the US-EPA), and integrated using agent technology.

The EDR is a single source of metadata information about the definition, source and location of environmental data. These metadata are information about data such as field name, field type, field length, value domains and system source. The TRS is a Web-accessible thesaurus of environmentally related terms, concepts, data dictionaries, and regulatory definitions. The TRS incorporates elements of a multilingual thesaurus, which will be used in EDEN-IW. These two reference systems provide a foundation for the construction of ontological reasoning systems.

Previous work, for instance by Lawrence Berkeley National Laboratory, has demonstrated the usefulness of the Extended Markup Language (XML) in integrating information from various sources.

Extended Markup Language is an enabling technology that separates content from formatting to facilitate the retrieval, publication and exchange of environmental data and metadata from databases, and has been extended to OWL and RDF. Although the use of XML does not by itself address the meaning of the data in an application, the recent combined initiative by World Wide Web Consortium (W3C) and DARPA to produce a more agent-oriented markup language, DAML (DARPA Agent Markup Language), is a step in this direction.

The previous work thus demonstrated the powerful technology, addressed by combining the use of:

- *Agents* – co-operating specialised objects distributed across a network, intranet, or Internet that reason about concepts, respond to queries, achieve goals, and detect complex events within the system.
- *Semantic brokering* – dynamic incorporation of new agents into the system and identification of agents that can provide desired services or data, based on semantic knowledge about their capabilities.
- *Ontologies* – specification of domain-specific vocabularies to represent real-world concepts and their relationships.
- *Multi-level task and query planning* – separation of the task plan requested by the user from the determination of how to process a specific query.

These features, combined as they are in distributed semantic agents, allow for a flexible, plug-and-play approach to integrated query processing and data analysis that would require many developer-years of effort to duplicate.

Some of the ideas developed in the hybrid agent multi-database system have been addressed in academic database and multi-database research projects in the past: for instance, the use of agents to accomplish these goals in a dynamic environment, integrated standard Web meta-language and metadata models, and open agent technology and relational databases in a larger, more distributed environmental application. Forward-looking developments could include incorporating an important multilingual thesaurus into the application, formalisation of value-mapping techniques using ISO 11179 metadata registries, enhancement of distributed hybrid semantic agents, and the development of reasoning engines that can take advantage of the newly developed XML data transport layer of Distributed Hybrid Semantic Agents in order to explain the occurrence of retrieval problems such as missing resources and absence or mismatch of data, as well as to identify the agent or component responsible for a slow response.

Fortunately, it is possible to address a part of these problems with state-of-the-art multi-agent technology, by employing multiple co-operative intelligent agents to conduct concept-based searches of heterogeneous, distributed information in a dynamically changing environment of databases and the World Wide Web. This type of technology can help organisations retrieve relevant information that currently is not easily accessed, filtered and retrieved by Web browsers. It may also help to harmonise the knowledge held in the databases, while maintaining the autonomy of data custodians. This multi-agent technology can be thought as dynamic virtual data warehousing, where physically separate databases are logically integrated into an efficient tool for decision support by a shared knowledge base. An overview of the technology is given in the next section, and described in detail in Chapter 4. The related security issues are outlined in Chapter 6.

1.3 TECHNICAL BACKGROUND

The objective of the EDEN-IW project is to make existing distributed environmental data available to citizens and other users through an intelligent interface acting as a one-stop shop. The users, who may also be public authorities (for instance, environmental regulatory agencies) and the public, will be able to address their needs for Inland Water data through one common interface, independent of physical or logical location of the databases, the database languages used, or of the specific nomenclature used in a specific database, and without knowing which database or databases contain the relevant information, by data mining the most complete set of information sources.

1.3.1 Agent Technology

The technology explored was the use of independent software agents. A pilot implementation is described in Haastrup and Würtz (2001). Each of the agents has a specific task in the complex process leading from the question raised by the user, through the formulation of specific (but database-independent) queries, through to the specific queries sent to databases which the agents evaluate as potentially having an answer or part of the answer for the actual question. Such requests for information require a common "language"—i.e. a list of accepted and well-defined words–which is the basis for an ontology, relevant for inland waters.

While data exchange for inland water, by definition, is centred on a regional level, within a common watershed, significant benefits can be reaped, by exploring the issues in a broader context.

Besides data collated directly by the users, potentially useful information is increasingly becoming available on the Internet on a global scale by a variety of information providers. However, the increase in quality and quantity of environmental data and information on the Internet highlights the difficulties of accomplishing effective and efficient retrieval. Terminology and language differences are two major barriers to information access and dissemination. Information services to the public, industry and environmental program staff can be greatly enhanced by deploying emerging Internet technologies that draw upon new semantic management techniques and tools based on types of multi-agent systems that communicate using an agent communication language based on speech acts and explicit ontologies.

Agent systems differ from competing database integration technology because they make use of intelligent agents to provide a flexible and dynamic information environment in which the arrival or departure of an agent only affects the availability of a particular service or set of data; such dynamic events do not affect the correct performance of the system as a whole.

The applied agent system, unlike most other agent systems, focuses on the use of a standard Agent Communication Language (ACL), which makes a seamless interlink and control of existing database technology possible at a much richer semantic level. The layout of the implemented agent system is illustrated in Figure 1.2.

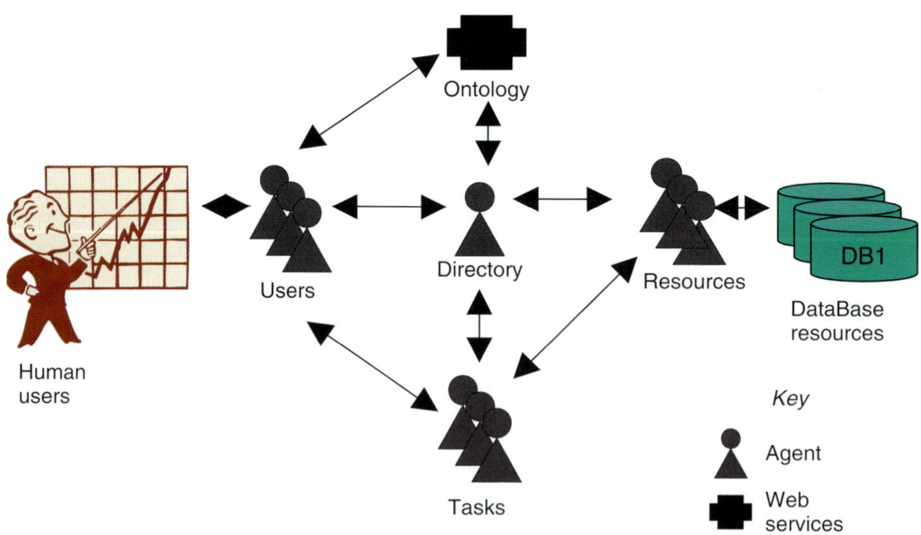

Fig. 1.2. Structure of the agent system in EDEN-IW.

1.3.2 Inland Water Databases

The aforementioned issues are relevant to a number of environmental areas, and certainly to the area of water management, where the rivers connect the inland waters among the various member countries, and where the costal and marine environment is shared.

On the issue of data, *data gathering and data mining*, the technical challenge is to be able to provide the necessary overview of the various data and information, while leaving the data in the original databases. This should happen in real time, and on-line, so that laborious transformation is avoided as much as possible. Since the databases are independent and geographically distributed, on different hardware and software platforms, issues of heterogeneity must be addressed.

Data on inland water quality and quantity is at present available primarily in databases only accessible for internal purposes in the national environmental ministries. Therefore, the users have problems in compiling data on Europe's inland waters, as easily accessible data are scarce.

Attempts to put all the available databases on the Internet through a one-stop shop are under way, but an intelligent query interface for the executive user or for the citizen is not yet operational.

Since the mid-1980s, significant improvement in data comparability on the European level has been obtained, through the regulatory process, with regular international monitoring and inter-comparison and with the improvement of analytical methods. The Internet means that the data collected are now easier to share, and that the data are compared with similar data, measured by others.

This tendency improves the data comparability with time. However, issues still remain on a number of different levels. One example would make this clear: traditionally, water quality has been measured on a number of parameters linked to eutrophication such as phosphate, nitrate, and chemical oxygen demand. However, in some European countries, these measurements are traditionally made on a filtered water sample, while in other countries they are performed on the raw sample, including the suspended solids. When the suspended solid content is low, these measurements may be comparable – while in the opposite case, they may be different.

In addition to the issues of environmental data and comparability, it is clear that a key issue is that environmental managers have to have ready access to concise information and data, in order to base their decisions on the best possible information and knowledge about the environment.

Both EU directives and national laws are paving the way for public access to this data, and some countries are experimenting with Web access to the data, though this is still uncommon.

Recently, various European organisations concerned with water have built databases on inland water issues, but a common interface and query possibility into their data are not available, with databases distributed all over Europe.

Let us examine an example of an organisation with concerns about water pollution. This organisation needs access to detailed information about the state of the environment in order to manage the situation. Rather than searching through many databases for information about this and where it may be located on government property, a user of the EDEN system would submit a question as, "Where in Europe do pesticides potentially represent a risk for surface waters?" The system would present the user with a set of graphical query interfaces and modes of interaction by placing a Java applet in the user's standard browser. To specify a query or question, the user could make a selection from the query lists. These are stored in pre-designed queries that offer selections of content such as time, place and pollutant. The system will then search for the information in a specified number of geographically dispersed databases. The results of the queries would be presented in simple text and/or graphical displays. The user is not required to be a database expert to know where the databases are, or even to know in which database(s) the data resides. These aspects are further discussed in Chapter 5.

1.3.3 Semantic Management

As has already been briefly discussed, a series of issues exists relative to the one of different languages, and, indeed, different concepts in the same language. These issues all relate to semantic management, where the key idea is to share data "content" freed from the incompatibilities of the technology platforms used to store and manage the data. It thus brings a focus on the data content, particularly on expression of the data semantics, such that compatible data can be drawn together from dispersed sources. An ontology is used to express the concepts found in the data, so that all sources can advertise their data resources in a common lingua. This project draws upon content and technologies used for semantics management as a basis for ontological development and also for mapping and integrating query results from disparate sources into compatible semantic equivalents that result in a uniform query result. This engages the semantic content for the demonstration system, the semantics management technology, and international standards under development for metadata registries and other semantic management technologies and techniques.

Previous work in the area of semantic management has been mostly focused on a multilingual environmental thesaurus. While this application will contain a thesaurus of environmental terms, it is better thought of as a thesaurus of carefully defined environmental concepts. Since the year 2000, a core of 6,000 well-defined environmental concepts have been created, giving each concept linguistic expression in 10 languages. This is described in detail in Chapter 10.

The power and usability of the thesaurus lies in the linkages of the many word expressions to a single concept and in its ability, through a hierarchical arrangement, to put those concepts into context. The thesaurus also enables a thematic view that cross-cuts the hierarchies. This allows relationships to be displayed around concepts. The concepts are clustered around some 40 environmental themes, such as air, water or pollution.

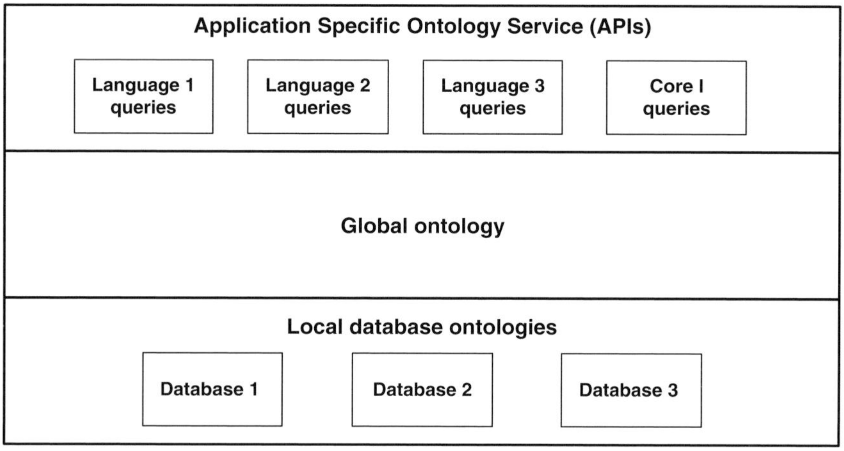

Fig. 1.3. The three levels of ontologies applied.

It is clear that such cluster of concepts have to be applied at the appropriate level. For the present discussion, the following three levels can be applied:

1. External language (for the users)
2. Common reference data language and
3. Specific data language of a database.

This can be shown schematically as in Figure 1.3.

1.4 SOCIAL CONSIDERATIONS

There is an increasing need to take the requisite measures to protect the environment, based on information that is accurate, timely and reliable. One of the main problems that policy decision-makers encounter in meeting this need is the challenge of using the Internet as a major mechanism to organise, access and analyse environmental information contained in online databases.

The data necessary to make decisions are geographically dispersed and heterogeneous, and the databases (systems) are often incompatible with each other. Without a new technology to draw data together to provide more comprehensive approaches to data access, retrieval, process, and analysis, the databases remain as they are now: geographically separated and only useful in isolation.

A key social goal is thus to develop a combination of essential elements to access and retrieve information contained in this case in Inland Water databases, in such a way that the Regulatory Agencies could optimise their mandatory tasks of assessing the quality of water, identifying trends in terms of improvements or declines and producing water quality indicators.

The need is clear for providing users – including environmental managers, regulatory agencies and the citizens – with better, more detailed, high resolution, and more comprehensive data and knowledge about the inland waters. This more efficient flow of information could lead to a better management of the environment, of the water resources available, and to improved conservation.

In addition, users like the EEA and the U.S. Environmental Protection Agency (EPA) have recently strengthened their commitment to provide ready access to information, utilising the Internet as a major mechanism for information sharing and dissemination. EDEN-IW can expand to ongoing efforts to improve information access and sharing with agencies located outside Europe.

Besides data collated directly by the users, potentially useful information is increasingly available on the Internet from a variety of information providers. However, as the quality and quantity of environmental data and information on the Internet increases, the difficulties of accomplishing effective and efficient retrieval are becoming more and more evident.

Terminology and language differences are two major barriers to information access and dissemination. Information services to the public, industry and environmental program staff can be greatly enhanced by deploying emerging Internet technologies that draw upon new semantic management techniques and tools such as the JADE and the ontology mechanisms.

Improving the management of inland waters will possibly impact the citizen in many ways, ranging from leisure, tourism, through fishing (both commercially and as a sport) to a direct impact on the agriculture and on the industrial users. The social and health benefits of clean drinking water are tremendous, and just minor improvements to the overall efficiency and effectiveness of environmental management and planning in this area can have a significant positive impact.

Important social goals include the dissemination of information to public authorities and through them to the European citizens. This flow of information should lead to a better-informed public, and to the participation of the public in the political process. This will also lead to a more efficient and effective management of inland waters on the local level, and through the national and international authorities, to a better water management on the regional and international scale.

The ability to compare data from different databases will be a powerful aid in decision-making and analysis, enabling scientists and remediation program managers to detect trends that might not be visible when data from individual databases are viewed in isolation. Potential difficult situations may be detected early, increasing the efficiency of the interventions, and a more sustainable use of surface water may be encouraged.

The fact that the data may ultimately be available to the common citizen will make room for more participation in the decision-making, leading to a higher degree of self-regulation and improvements on the local level as well.

1.5 WATER ISSUES IN EUROPE

When the EDEN-IW project was conceived in early 1999, the Water Framework Directive had not yet been adopted.

The Directive was on its way through the legislative system, and, at that time, it was not yet possible to foresee neither the adoption date nor the development of the details of the Directive. Now, in 2006, the picture that emerges is that the Water Framework Directive has shaped the regulatory landscape in the European Union, to such an extent that it is the single most important piece of legislation in the water area.

Daughter directives are on the way, old legislation is expected to be incorporated or removed, and the decision-making processes linked to the Water Framework Directive are in place, with a Common Implementation Strategy for all Member States.

At the same time, the Water Framework Directive requires a significant scientific and technical input in the process, in order to arrive at the goals of the directive, which can be stated simply as "good ecological status" for Europe's water bodies by 2027.

In the practical implementation of the Water Framework Directive, the Joint Research Centre (JRC) is active in hosting the Water Information System for Europe – the information system which now receives and contains the Member States' data as they become available. This will be discussed in detail in the following chapters.

1.5.1 Paradigm Shift to River Basins

The central paradigm shift in the management of the water resources, introduced on the European level by the Water Framework Directive, was the shift from the management of water resources according to administrative boundaries to the management on the catchment, water-shed, or river basin level (Figure 1.4).

National boundaries will obviously continue to have an impact, but the real management has to follow the natural geographical boundaries of each river basin.

Thus, environmental managers in the water area have had to exchange their normal administrative map of Europe with national, federal, provincial and other administrative boundaries, for a map of the river catchments in Europe.

These issues are further examined in Part III of this book.

1.6 BOOK OUTLINE

Starting from the above mentioned needs and technological challenges, Part I of the book examines data exchange issues and previously used solutions. First the current overview chapter highlights the issues in Environmental Data Exchange for Inland Waters – issues which are common to most countries and especially to river basins, that are managed by different authorities. In fact the issues are similar, independent of whether the managing authority is shared among different local authorities, shared between federal states or even shared between nations. The issues of surface water management (and to some degree also of ground water resources) follow the relevant water basins – which are (still) given by nature.

For Europe today, the map of the river basins seen in figure 1.4 illustrates the situation. As seen the river basin boarders do not generally coincide with the national borders. This means that data exchange between national states are a necessity, in order to make a coherent overview of the total water management needs.

From this more general perspective, focus is shifted to an example of a national system in a large European country, France, where the data needs to be aggregated to the national level, based on a number of local authorities.

From this a more general discussion about how the thematic reporting by various authorities in the past, can evolve into a shared information system, which satisfies the necessary requisites for providing integrated data and information to the policy makers and the public.

After thus having set the scene from the water management point of view, three chapters (chapter 4 to 6) initiates the discussion about the information technologies, which can be used to facilitate the data exchange in the environmental area.

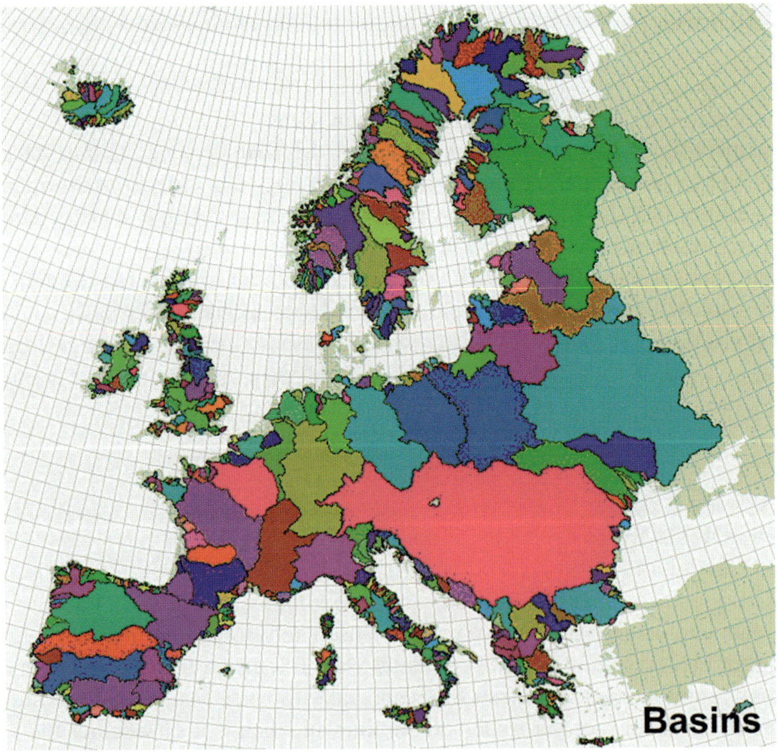

Fig. 1.4. The river basins in Europe.

First, in chapter 4, a detailed description of the issues linked to the use of software based on the "independent agent" concept is made. Such software agents have the potential to react in a more intelligent fashion, than was previously possible to envisage with information technology. Chapter 4 describes some of the most pertinent earlier developments. From this technology view, attention is then shifted to the various options for sharing of database data, and for obtaining database integration. A significant number of technologies are in existence, each based on different views of the future information technology options.

Finally, in chapter 6, the various security issues, in a distributed world, are examined in general. These security issues are becoming more and more in focus over time, and carry the well known dilemma of finding the right balance between having a secure, safe and full integrity system and a system with full openness to the outside world. These two states are the extremes in a spectrum, where each data provider needs to find an acceptable compromise between complete closure to the outside world (which gives a good guarantee for security, integrity and safety, but precludes data exchange), and the full openness.

From this general setting of the scenes in Part I of the book, focus is then narrowed in Part II to the results of the Environmental Data Exchange Network for Inland Water project (EDEN-IW). It was a three year research project, financed by the IST programme of the European Commission under the 5th Framework Programme. It started officially in 2001 and came to an end at the end of 2004. Initially the project is introduced and an overview of the results are given in chapter 7. Chapter 8 then focuses on the results obtained through the use of multiple independent software agents in a distributed system, and especially in systems with distributed databases. The research undertaken showed the needs, the power and the usefulness of ontologies, where various domain specific keywords are related to each other in such a way, that meaningful inferences can be made. This is described in chapter 9. Part of the work on ontologies is also related to the – for Europe so important – issues of multiple languages. A major undertaking in the project was linked to this area, and to producing the relevant terminology reference system for the project and for further use in the future. This is described in chapter 10. Three chapters (Chapter 11, 12 and 13) then examine results from the project, and from the pilot implementation made. First the technologies for visualizing agent communication (Chapter 11) are examined, followed by a more detailed description of the user interface and the possibilities for using the system (Chapter 12). The needs for geographical information and map services are then examined in Chapter 13. Finally the more detailed examination of the results of the project is supplemented by a discussion about performance issues of such distributed system.

Finally Part III, examines data harmonisation issues and perspectives, and looks towards the future, with contributions related to the collaboration across the Atlantic and the drive towards a common water Geographical Information System (GIS) in Europe. This starts from a more detailed discussion of the needs for Environmental data, the limits of current technology and the current position of the research edge. After this an examination of harmonisation issues for the basic data for water in Europe, based on a modeling approach and an underlying digital elevation model is described. For such a model approach it is assumed that water is running downstream under the influence of gravity, which is only partially true, in the well regulated waterways of low lying countries. In for instance the Netherlands, the water flow is heavily influences by man through the use of pumps and active management by opening and closing of dams and dikes.

Finally, Chapter 17, describes future options and perspectives for the data exchange in new and coming information systems for water at the European level.

Chapter 2

AN EXAMPLE OF A DATA EXCHANGE NETWORK: THE FRENCH SYSTEM FOR WATER DATA

D. Preux

2.1 THE FRENCH SYSTEM FOR WATER

A National System for Information on Water (SIE) (French Parliament, 1992) aiming at providing access to information scattered between several servers has been operational in France since 1992. The new European regulations will bring a new impetus to this initiative, which has more than 10 years of experience.

2.1.1 A Need for Unified Access to Water Data

The idea of National Information System on Water is an old one, and has its origins in the management principles developed for river basins following the French water laws of 1964 and 1992.

For integrated management to work, water organisations and the public must share the same information in order to make management decisions. This is an ambitious aim taking into consideration the large number of independent organisations which are active in the area of water. For instance, at least 6 ministries have responsibilities linked to water, and they are active within their central government areas and also have widespread services throughout the 99 administrative areas of the French government system.

Added to this are the water agencies of each of the six main river basins, research centres, and the public establishments with responsibility for underground water, shorelines, etc., plus all the local organisations at municipal level.

Finally, add the large water companies, which play an important role in France. By way of example, the surveillance of river water alone involves 157 different producers of data.

In order to be fully useful, access to current water data must be accompanied by access to historical data, as environmental events are often slow to develop. In this way a system for water information cannot merely be restricted to the question of access alone, but must also be concerned with the preservation of data in the long term within a framework of permanent data banks.

In 1992, the National Network for Water Data (RNDE) was set up to create an information system coherent to all its users. The word "network" is used on purpose, and it reflects the partnership element within the SIE.

This initiative was started long before the Internet was widespread, and it delivered specific information to project managers within their area of responsibility. It created a shared semantic structure and reference system, and large data banks organised by topics.

2.1.2 *Responsibilities*

This is an area of great importance and poses several important questions.

Data quality
How is it possible to know how high the quality of data is, and to inform users of this? Quality is a relative notion since it is measured in the use to which the data will be put ("fit for purpose"). For instance, in hydrometry, flow measurements designed for tripping flood warnings do not have the same precision as those aimed at overall, deep knowledge about the water flows, which need to take into account small and apparently insignificant changes, such as those caused by climatic changes.

The responsibilities of participants
What are the responsibilities of the data producer and the operator of a data bank when false, erroneous, or low quality data is sent out to users?

The rights of data users
Do users merely have the right to permit consultation of the data, or may they reuse, develop, and even commercialise the data used?

2.2 INFORMATION SYSTEM STRUCTURE

The French system of water information has not sought to centralise the data within large databases, favouring distributed systems.

However, with the available state-of-the-art technologies at the start of the 1990s, it was difficult to construct distributed databases. It became necessary to create several large databases so that the most frequently accessed data is available without (severe) delays.

For this reason, a number of national data banks were set up: the BNDE, for river water quality; the PLUVIO, for rainwater; the HYDRO data banks, for flow in watercourses; the ADES, for level and quality of water in aquifers; and the BHP data banks, for the condition of fish populations. The data banks were connected together in the SIE system. This is shown in Figure 2.1.

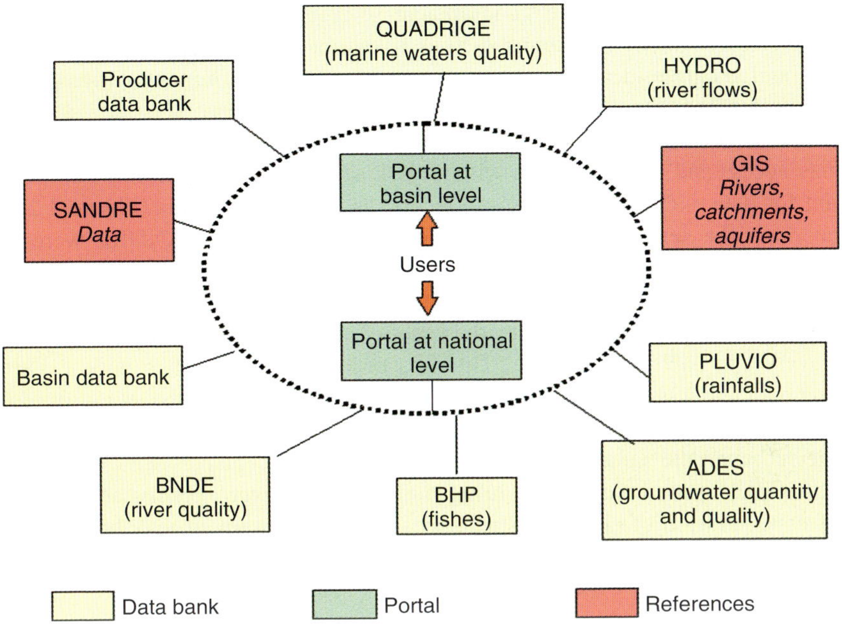

Fig. 2.1. Architecture of the French SIE information access system.

The site design includes a portal giving the data available (metadata) and guiding the user towards the data banks (or, in case of failing that, to the data producers) in order to access the data.

The databank for river quality, BNDE, was included in the EDEN-IW project.

2.2.1 Semantics and References: The SANDRE Module

It quickly became apparent that in assembling data from several different sources, even those dealing with the same subject (for example, water quality analysis), there were problems in terms of basic concepts. The consequences of this can be severe, when (without knowing it) comparisons are made between different sets of information that are not comparable.

The construction of a national reference model became the starting point for a priority project named SANDRE: "Service d'Administration Nationale des Données et Referentiels de l'Eau" (National Service for Water Data and References). SANDRE has the same role as the "global ontology" in the EDEN-IW project. It is concerned with all areas of water, and is not restricted to the quality of waterways (SANDRE, 2005).

With the aim of facilitating shared use of information from a variety of sources, it is essential to be able to rely upon a single system of reference that is used by all. This complements the semantic work of SANDRE, setting up national codes for substances, analytical methods, etc.

In the same way, large-scale geographical references are required to describe and codify in a single, unified way items such as rivers, lakes, underground water, etc. Then a positive and accurate identification of the location of measurement stations, sampling points, or polluting sources can be established.

The mission of SANDRE is to define water data in a highly precise manner so that all users may access and exchange data without risk of errors. SANDRE has thus become a key point for data exchange between producers, databases, and users (Figure 2.2).

Concerning the semantic elements, the data dictionaries of SANDRE were created in the formalism of entity/relationships data models. The exchange formats which were originally set up for EDI message formats have since been converted to XML (Figure 2.3).

Building a data dictionary requires the input of a large range of practitioners in order to take every approach into account. This is a task which demands a wide expertise in modelling data and a good knowledge and understanding of the elements involved (water quality, de-pollution plants, etc.), in order to be able to identify the common elements and to detect when the same term is used for different concepts.

For example, in the case of measuring the quality of rivers, results are expressed in terms of concentration of chemical substances. This measurement may have been taken in the water itself, or in filtered water, or on particles in suspension in the water, or on algae, or on sediment, or even in living creatures such as fish. Therefore, the concepts must take into account the medium (plain water, material in suspension, algae, sediment, fish) and, for each medium, the analytical fraction (filtered water, fish stomachs, etc.) (Figure 2.4).

The measurements of the same substance in different media and/or in different analytical fractions are not comparable with each other, especially as some may represent instant pollution (in water), accumulations within a few days or months (fish) and even longer periods (sediment).

It is clear that the apparently simple notion of measuring water quality reveals a far greater complexity.

This complexity must be correctly analysed and modelled at the technical level, otherwise grave errors of interpretation are likely and must be invisible for end users.

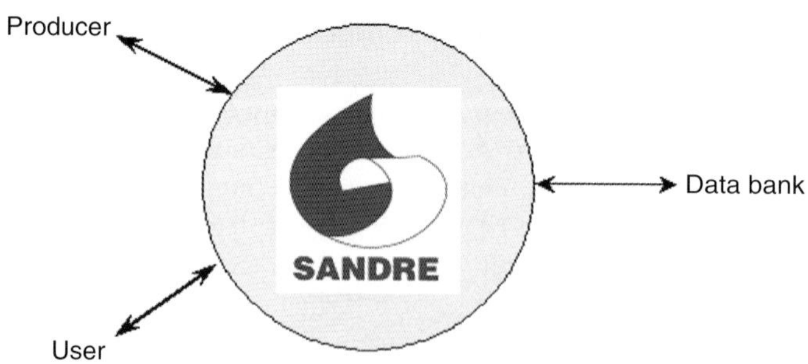

Fig. 2.2. SANDRE, at the core of exchange of data.

```
<?xml version="1.0" encoding= "UTF-8"?>
<LABO_DEST xmlns=http://www.rnde.tm.fr/francais/sa/xml/scenario/labo_dest/1>
<StationPrelevement>
    <CdStationPrelevement schemeAgencyID="1">06047200</CdStationPrelevement>
    <LbStationPrelevement>Reyssouze à Pont de Vaux</LbStationPrelevement>
    <Prelevement>
        <DateDebutPrel>2004-03-20</DateDebutPrel>
        <CdPrelevementCommanditaire>06047200-01012005-PESEAU</CdPrelevementCommanditaire>
        <Support>
            <CdSupport>3</CdSupport>
            <LbSupport>Eau</LbSupport>
        </Support>
        <Analyse>
                <NoAna>1</NoAna>
                <RsAna>0.020</RsAna>
                <RqAna>2</RqAna>
                <LqAna>0.020</LqAna>
                <InsituAna>2</InsituAna>
                </Parametre>
                    <CdParametre>1083</CdParametre>
                </Parametre>
                <FractionAnalysee>
                    <CdFractionAnalysee>23</CdFractionAnalysee>
                </FractionAnalysee>
                <Methode>
                    <CdMethode>386</CdMethode>
                </Methode>
                <UniteReference>
                    <CdUniteReference>133</CdUniteReference>
                </UniteReference>
        </Analyse>
</StationPrelevement>
```


Fig. 2.3. Extract of river quality data converted to XML format.

This must be borne in mind when planning the uses which require heterogeneous access to information. On the one hand, the construction of a "Global ontology" is a major task (to a large extent independent of technology), and on the other hand great efforts still have to be made to produce useable information for decision-makers and (from simplified data) for the public.

2.3 PERSPECTIVES

For two years, the French system of water-related information has been the object of profound reorganisation, in order to take into account new demands for water regulation

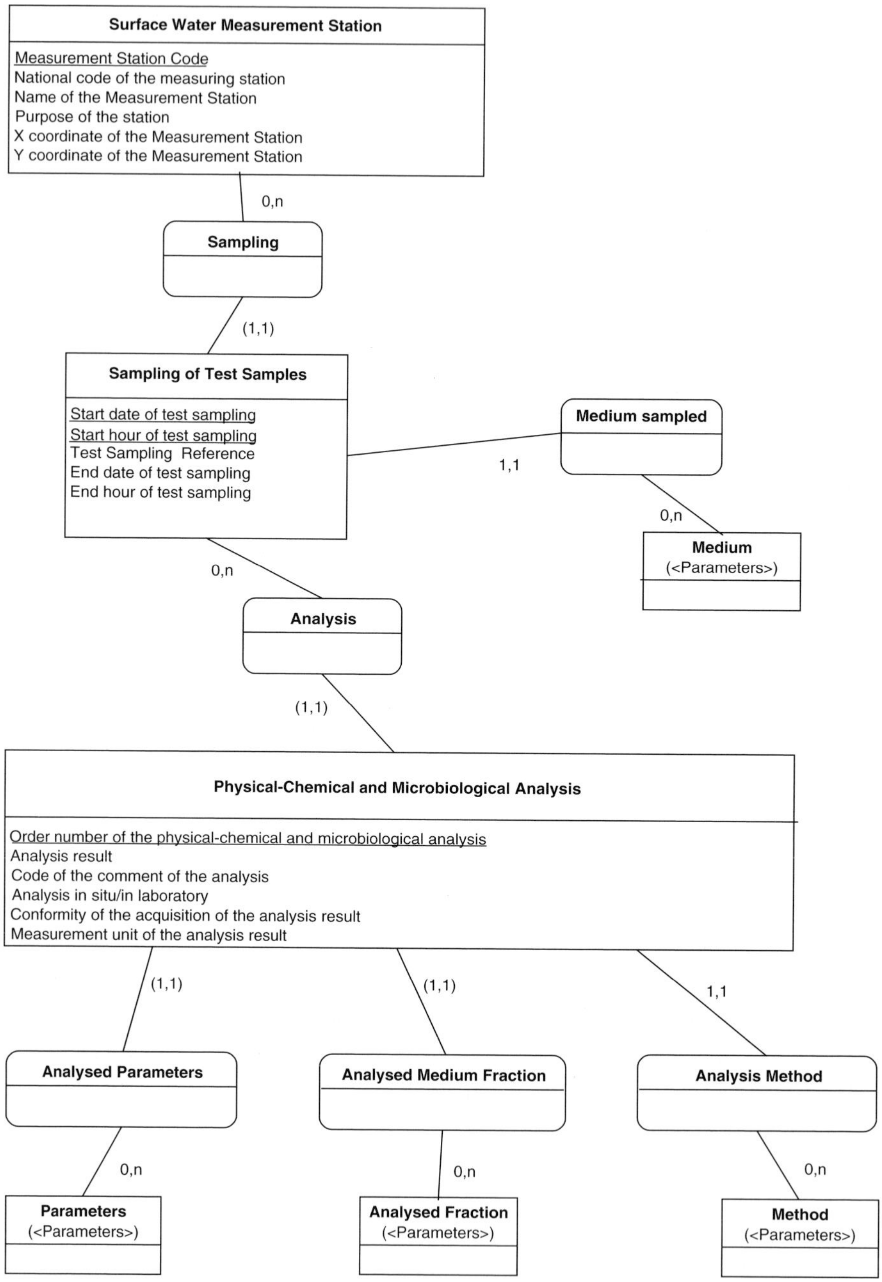

Fig. 2.4. Simplified SANDRE model of measurements of river water quality. (The text has been translated. Indeed the actual system is in the French language).

and access to data, and to take into account the opportunities created by the latest technology.

2.3.1 Regulation

The *European Water Framework Directive* (EU Parliament & Council, 2000), which came into effect at the end of 2000, requires the introduction of a far greater information system than any used previously. Its demands include not only tracking the condition of the environment, but also compiling an inventory of the human pressures and to monitor their impact. It requires an economic analysis for checking cost recovery plus a cost–benefit analysis to justify measures. The Water Framework Directive also imposes a duty to inform the public, with a view to its participation in decision-making. These demands clearly imply the production of more data, including in areas where little study has previously been made, and to publish them to the lay public in order to facilitate involvement in decisions.

The *Aarhus convention* (UNECE, 1998), which is applicable in Europe, imposes the obligation to provide public with access to almost all environmental information. In practice, all data on water are included, and in order to limit the cost and workload for the responsible bodies, this access must be automated.

The *Directive of January 2003 concerning public access to information* (EU Parliament & Council, 2003) on environmental matters makes the various States responsible for holding metadata files.

The project of *European INSPIRE Directive* (EU Parliament & Council, 2004) is concerned with geographical information, especially on environmental themes. It will impose the setting up of servers carrying metadata, with the capacity to interoperate with other geographical databases.

2.3.2 Technological Development

Widespread use of the Internet brings a new perspective to information systems. Data modelling tools (UML, RDF, OWL, etc.) and especially exchange tools (XML, GML) are now available. Furthermore, a lot of material is available for metadata, e.g. Dublin core and ISO 19115.

2.3.3 The Water Information System in the Future

The water information system is confronted with a threefold challenge of managing more data, improving information access, and making the various systems interoperable.

It must also take into account the obligation to communicate with the WISE (Water Information System for Europe), being implemented on a Europe-wide basis under the Commission.

Thus, the system must be decentralised, must include a greater quantity of data, and be able to provide a means of access for users who are not concerned with storage or formatting.

A calendar has been defined to upgrade the present water information system:

- *End of 2005*: Selection of the architectural principles and technical means. Prototype testing to validate the selection.
- *2005–2006*: Construction of a national portal with a numerical and geographical interface.
- *2006–2007*: Upgrading of existing databases to be connected to the system and creation of new databases.

2.4 TECHNICAL PRINCIPLES AND CHOICES

Regarding the technical principles and choices, a "Green Book", which sets out the broad principles to be followed by the future French Water Information System (SIE) has been drafted (French Ministry of Ecology and Sustainable Development [Lalement and Lagarde, 2005]). The main aspects, whose scope renders them applicable to any form of system to access disparate information, are set out below.

2.4.1 Dispersed Architecture and Subsidiarity

The National System for Information on Water relies in part on the existing IT infrastructures amongst the partner members, who establish these systems for their own purposes, and with different needs. SIE does not seek to unify, and has, therefore, adopted a heterogeneous, dispersed architecture, favouring the inclusion of existing methods created for the members' own purposes. The common infrastructures managed by SIE do not include functions better managed at the partners' local level.

Each partner retains complete freedom to apply other means of access to their data, other than SIE, as well as with SIE.

The System is planned not to constrain a partner to certain technical choices made on a central level, and a partner may refuse to put in place any sub-system to be integrated with SIE, if this is likely to affect the performance or security of their own system.

2.4.2 Data Unity

In order to simplify the data producers' tasks, they are asked to place their data in just one database or reference bank. This rule ensures that the inconsistencies, which arise as soon as a piece of data is stored at two different locations, are avoided. This principle of data unity does not rule out technical replication, which may be justified on the grounds of performance or cost.

2.4.3 Semantic Interoperability

The National System for Information on Water relies on two levels of semantic interoperability:

1. A conceptual level, which allows a common understanding of objects and relationships between them (for example, the definition of a monitoring station or how a sample is linked to a monitoring station). These are the dictionaries established by SANDRE in UML format, XML schemes, RDF graphs, or OWL ontologies.
2. A reference level, which allows a single identification system (code or label) to be shared by users. SANDRE defined the sources to be able to attribute codes (e.g. ISO3166 for countries, INSEE for local authorities, SIRET for companies, etc.). Where no appropriate authority exists for codification, (as in the case of water analysis parameters), SANDRE is responsible.

2.4.4 Technical Interoperability

The technical interoperability rests on current norms or de facto standards, such as PNG (Portable Network Graphics) or XML, HTTP, SMTP (Simple Mail Transfer Protocol), SOAP (Simple Object Access Protocol), etc.

The original data banks may take any physical or logical format. It is only the data exchanges that require a single format to be employed.

2.4.5 Open Standards

The choice of open standard software and exchange formats ensures the durability of information. SIE relies on standards, with respect to the following conditions:

- They are to be defined within a framework of a participatory decision making process, accessible to all at an affordable price.
- They must not be suborned for use for commercial purposes.
- There must be at least one application within the norms, which does not require additional access rights. The Norms containing these conditions include: IETF (Internet Engineering Task Force), W3C (World Wide Web Consortium), OGC (Open Geospatial Consortium), DCMI (Dublin Core Metadata Initiative), OMG (Object Management Group), OASIS (Organization for the Advancement of Structured Information Standards), JCP (Java Community Process).

2.4.6 Use of GIS

Use of maps is fundamental to environmental information, as the majority of data are georeferenced. SIE normally uses GIS according to the principles set out in INSPIRE (the EU initiative for creating a common spatial data infrastructure) and ensures full access to geographical information.

2.4.7 *Access to Data*

The National System for Information on Water identifies two types of data: visual consultation data (e.g. PDF documents) and data for use (e.g. downloadable and reusable as with XML files or text files with separator).

In theory, all elementary data must be able to be used, all synthesis data must be consultable, and metadata must be available for both consultation and use.

2.4.8 *Metadata*

Metadata are essential to inform users about the existence of elementary data, synthesis data, or services. SIE manages the metadata catalogue.

The following formats are recommended:

- WSDL (Web Services Description Language) 1.1 for Web services
- Dublin Core for HTML and XML documents
- ISO 19115 for geographical data
- RDF for Website descriptions.

2.4.9 *Web Architecture*

The Web corresponds to the SIE architecture needs according to the schematic in Figure 2.6. A range of technologies are to be tested and evaluated. Solutions provided by Web services (call to distant procedures, dispersed objects, XML messages) are preferred:

- REST approach based on HHTP/GET and POST with XML message exchange, better suited to search and consultation of data.
- SOAP approach, based on http/SOAP and SOAP message exchange; better suited to insertions, modification, and deletion of data.

2.5 DISCUSSION

Environmental information systems (more than any other systems) need to group and connect existing heterogeneous systems, because of the large numbers of organisations involved in the management of environmental issues.

Additionally, the legitimate interests of citizens in environmental matters make it necessary (legislation even makes it obligatory) to give very wide access to data.

Environmental information systems, such as SIE in France, must therefore respond to this challenge by providing unified access to heterogeneous and distributed databases.

The National System for Information on Water has shown that matters of organisation and management of partnership projects, the rights of data use, and commonly understood and shared concepts are essential for such projects to work, and separate from any technical considerations.

The existence of SANDRE at the very heart of semantics aspects demonstrates that solving the problem of common language and reference is a long and laborious task.

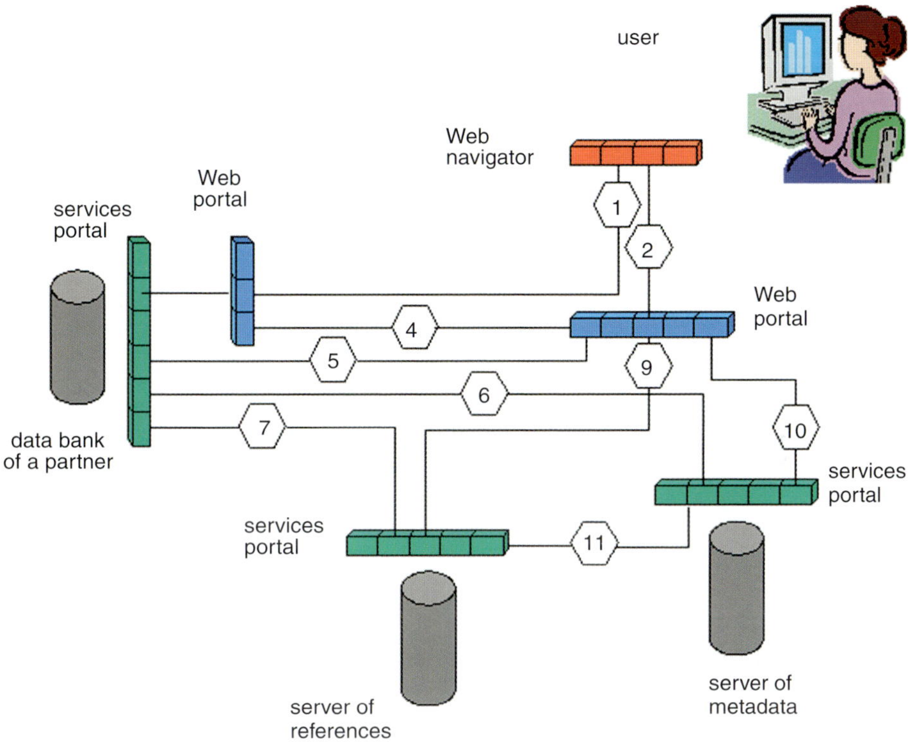

(2) A user accesses the SIE portal to find what resources are available (10), in order to consult a partner's database (5) or to consult references (9);
(4) SIE Website and the Website of the data bank are linked by a contents syndication
(5) SIE Website uses services supplied by the portal of the partner's Website, e.g. requests for database
(6) Database service portal uses the references metadata server (consultation, registration, and publication of metadata)
(7) Database service portal uses the services of the reference server (consultation and update of references)
(1) User retains the option of going directly to the database portal
(11) Connection between the reference server and the metadata server

Fig. 2.6. Principles of the Web architecture.

However, it does represent a very special trump card for SIE, as it has delivered the "global ontology", which is absolutely needed for this type of project.

In terms of a technological solution, SIE leans towards the approach of "Web service" kind, which is most readily recognised in the French information technology community.

Chapter 3

THE EVOLUTION OF THEMATIC REPORTING INTO A SHARED INFORMATION SYSTEM: AN EXAMPLE ON WATER

S. Jensen

3.1 HISTORIC LEGACY AND POLITICAL BACKGROUND

Environment policy making has evolved from targeting specific problems through directives aimed at curbing environmental pressures from point sources, such as dangerous substances from industry (e.g. in the 1970s and 1980s), to addressing sectoral influences on environmental pressures through broader-based approaches (e.g. urban waste water treatment, auto-oil programme in the 1990s), to addressing environmental problems through integrated management approaches (e.g. EU water framework directive, biodiversity).

The 6th Environment Action Program (EAP) thematic strategies continue this trend. Implications for information provision are that policy needs are changing and increasing, but the legal basis has not been adapted. For example, the reporting of information is still stuck, based on the perspectives of 1970s and 1980s. Hence, gaps between policy needs and data availability (on a legal basis) are increasingly growing. While there is a need to build on recent initiatives, it is necessary to think about some radical (yet cost-neutral) solutions to close the gap by 2010 and beyond.

This puts up new challenges to the policy arena as well as drawing a need for new approaches to our information systems.

The necessary political and organisational changes build on current and future initiatives on streamlining reporting and revised reporting directives.

Those initiatives are centred on working towards a shared information system comprising of common (spatial) data policies as well as new ways of production (monitoring and assessment) and dissemination of environmental information in support of EU policies.

In parallel, harmonised technical solutions need to be applied to support the sharing of environmental data and the challenges towards data integration, modelling, and assessment.

European Union's 6th EAP recognises the need to revise the reporting system in order to achieve a clearer specification of policy-relevant information needs to increase transparency and to remove overlaps and duplications. The programme highlights that solutions enabling a more transparent and efficient flow should be enabled, and that suitable IT developments should be exploited.

3.2 IDENTIFYING BUSINESS PROCESSES AND STREAMLINING REPORTING OBLIGATIONS

In order to improve knowledge on how environmental reporting is done to date, and what can be proposed for the future, it is understood that this is not solely about better technical developments (e.g. reporting tools) but first and foremost about processes which lead to reporting requirements and potentially to regular data exchange.

A European information system could cover the state of the environment as well as compliance, effectiveness of transposition, and practical implementation, then requirements, but would require a high level of commitment.

To achieve that, agreements on new working methods and on a more relaxed approach to published data in relation to compliance checking will, therefore, be needed at several levels. Compliance could be checked in new ways as well, if agreement is possible all over Europe.

One way of starting a "business process review" is to analyse the current obligations, and then to provide how a more systematic output of usable environmental data may be obtained.

The European Environment Agency (EEA), therefore, took a layered approach by identifying all the obligations around and then examining in detail those that constitute data flows.

This analysis of the reporting obligations stored in the EEA Reporting Obligations Database has supported the claim that member states are subject to a significant number of reporting obligations, partly derived from the directives of the European Union, and partly derived from various conventions signed by the member states.

The number of reporting obligations implies a heavy reporting burden. For example, EU countries like Germany, Sweden, Finland, or Denmark have more than 350 reporting obligations assigned.

The burden of reporting will also be dependent on the effort required to produce the requested data. Compliance reporting accounts for 41% (by number), making it the biggest fraction of reporting obligations. As this is less than half the total number of reporting obligations, the biggest reporting burden, in general, is perceived to originate from the state of the environment and policy effectiveness reporting where the effort in gathering the data to be reported is bigger.

Two-thirds of the reporting obligations relate to the environmental issues of biodiversity, nature, and water (and here mainly marine water). The split between compliance and other reporting for these issues suggests that the reporting burden would be bigger for the water issue. This finding is also backed by an examination of the number of parameters to be reported. Such a view on ROD (Reporting Obligations Database) data further suggests that the issues of air and climate change and chemicals could demand a similar workload to that of the biodiversity reporting.

Biodiversity and water are the areas with the biggest number of players, as these areas are the main focus of conventions with almost 90% of the reporting to conventions concentrated in these areas.

An examination of the reporting obligations with a view to identifying possible duplicate reporting (i.e. reporting of the same data more than once) based on the classification of obligations performed in the study suggests that the potential for such duplicate reporting is much bigger in the water area than in biodiversity.

The conventions where the EU as a whole has entered as a contract party along with individual member states represent another possibility for duplicate reporting, if the member states first had to report to the EU prior to the convention deadline (so that EU could create its aggregated data). Of these four conventions have been investigated, which would account for more than 75% of such reporting: OSPAR, Bern, AEWA, and Mediterranean Action Plan. In a more detailed examination of this, it has been found that this type of duplication is not the case for any of the four. It is estimated that this does not represent a general issue with respect to duplication of reporting.

There is a general trend that compliance reporting is more legally binding than the state of the environment reporting obligations, which are often moral or based on voluntary submissions of data. This may be of no real significance, however, since in the area of international law, effective policing is not an option.

Reporting obligations towards convention bodies are split roughly equally between compliance and state of the environment, whereas reporting towards EU – and especially the DG Environment – has a majority of compliance reporting, although it can be determined that the state of the environment reporting for DG Environment occurs at a higher frequency than compliance reporting does.

A correlation of the findings in this analysis with similar studies confirms these findings, and leads up to the general recommendation: to separate compliance and state of the environment as much as possible in favour of having international bodies – notably the EEA – to perform the effectiveness reporting based on a set of broadly agreed and accepted indicators.

The second level analysis will be referred to later, taking the thematic area water as a focus.

3.3 REPORTNET AS A CURRENT TECHNOLOGICAL SOLUTION TO STREAMLINE REPORTING

Since the year 2000, the EEA and the European Environment Information and Observation Network (EIONET) network, with financial support from the DG Enterprises' Interchange of Data between Administrations (IDA) programme, have developed a modern system for reporting.

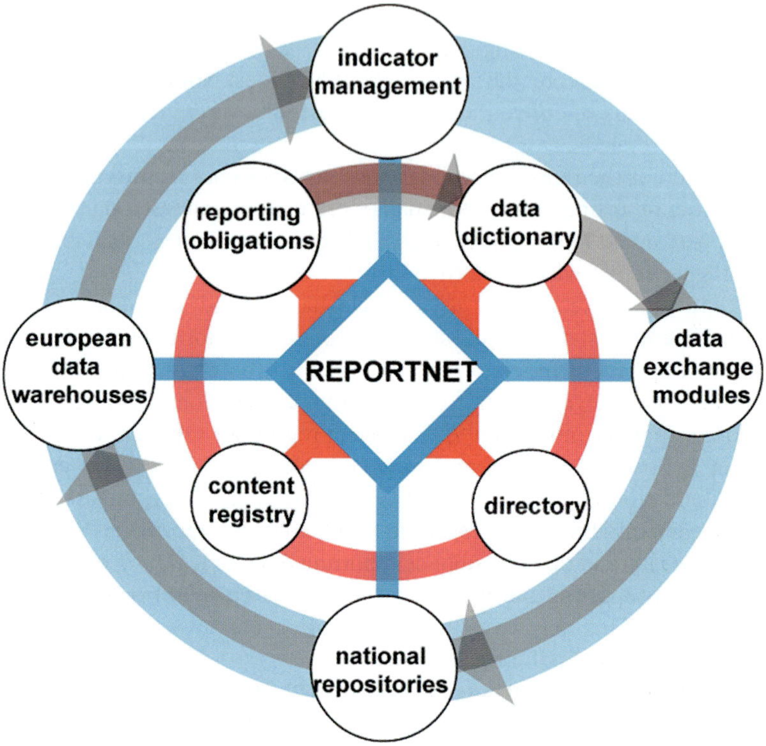

Fig. 3.1. The Reportnet set of functions including model data flows.

Reportnet is both an information infrastructure and a new setup of functions to assist data and information flows resulting from reporting obligations. It is currently mainly being used for reporting to EEA, hosting some of DG Environments' reporting tasks. The open system allows deliveries also to other national and international organisations.

Reportnet builds on active use of the World Wide Web. The system integrates different Web services and allows distributed responsibilities.

Reportnet ensures that the delivered data is made quickly available for analysis and, in addition, publicly available without delay if conditions allow for it (Figure 3.1).

It is a contribution to what is a shared challenge for European countries, their networks of institutions and EU institutions – to develop technical and organisational systems which facilitate streamlined, transparent, and shared use of multipurpose data and information.

3.3.1 The Principle of a System for Multiple Use: Deliver Once – Use by Many

Countries also deliver data to, and work with, Eurostat, DG Environment, other DGs, OECD, UN, and international conventions. These international institutions also have a shared interest in improving the delivery process and reducing countries' burden of delivering overlapping data with numerous different methods and tools.

In the future, it will be a joint challenge for the European countries and the international institutions to maintain a coherent infrastructure for information flows delivering multipurpose data files. EEA and the member states work together to have Reportnet contribute to this challenge. They are making data and information available, so as to maximise their multiple use and value. As a result, countries are able to deliver with higher response rates, more timely and better quality data for more numerous targets.

3.3.2 Functions Offered by Reportnet

Reportnet contributes in several ways to more effective data/information flows:

- It provides a Web catalogue of reporting obligations to the countries' planning. The Reporting Obligations Database (ROD) shows reporting obligations for each country (http://rod.eionet.eu.int).
- This catalogue can also act as a portal to necessary detailed information for those delivering data on a specific obligation. It has links to guidelines, technical specifications of data to be delivered, contact information of institutions and persons responsible for deliveries in countries, and to the underlying legislative acts or conventions.
- On selected main obligations, it can provide detailed specifications on data to be delivered, delivery templates in agreed formats, and guidelines (http://dd.eionet.eu.int).
- It provides smart software to make data provision by countries easier, more comparable, quality assured and fully audited (generic data exchange modules, GDEM, where necessary specialised data exchange modules for PCs, e.g. AQ-DEM and AE-DEM).
- It stores delivered data/information in one place for easy reference and for archiving over time (http://cdr.eionet.eu.int).
- It provides the means for tracking the status of deliveries on an obligation and accessing deliveries (http://cr.eionet.eu.int).
- It provides the means for other international organisations and countries to use other countries' information for their own assessment needs (CDR central data repository).

In the future, indicators shall determine more systematically what is needed to support the policy process. Therefore, a link to a more specialised indicator management system has recently been added (http://ims.eionet.eu.int).

Reportnet has been developed with funding from DG Enterprises. The development was done by taking on board the European eGovernment approaches, encouraged by the programme.

3.4 REVIEW OF MONITORING

Pressure is building up on monitoring systems in a time where the cost of environmental monitoring matters more, and where information gaps have still not been addressed and filled.

Monitoring systems in Europe are diverse and are not targeted towards the same needs. This creates problems if the aim is to build a European information system. We may have to study the possibilities for harmonising monitoring, to be able to meet the needs of new targets.

An additional challenge will come from the need to include modelling results more systematically, in order to fill the knowledge gaps. This is due to the increasing cost to provide suitable monitoring information from the present reporting schemes.

3.5 DISTRIBUTED INFORMATION SYSTEM

Future information systems will have to build much more on distributed technologies and shared responsibilities in an open network architecture where institutions can join in without giving up independence.

Services will become less centralised and will increasingly be provided by the European countries and their regions. The principle of subsidiarity will support the strengthening of this approach, and technologies will help to enable it.

Requirements will arise on the quality of data – and IT tools – in such a much more ambitious and targeted system. If everybody will have access to data and be able to aggregate, then agreements are needed on how to ensure a good enough quality.

Practical experiences have been gained to develop a distributed architecture for a good decade. This architecture started with the early usage of middle-layer technologies (CORBA, RMI), and evolved to apparently more intelligent solutions exploring agent technologies or web services. The challenge is still to upscale the existing examples.

In order to make these services combinable and compatible in an information system, more success on standardising environmental data and exchange formats is necessary. Where standardisation fails, cost will hamper the flow of information and its sharing and comparison. It is expected that the INSPIRE initiative will push these standardisation efforts forward with regard to spatial data, and that GMES (global monitoring for environment and security) will improve the monitoring side by providing guidance funding for projects by steering research in this direction, thereby filling knowledge gaps.

3.6 FROM WATER DATA FLOWS TO A WATER INFORMATION SYSTEM

3.6.1 Streamlining Obligations and Data Flows

The way the area of water has developed over the years is a very good example of how the processes and technologies are maturing and modernising. Therefore, it is a good idea to make a systematic study on how the issues raised in the previous chapters influence these developments.

This analysis of the data flows associated with the reporting obligations leads to the conclusion that the obligations for the water area are less scattered in the water area than

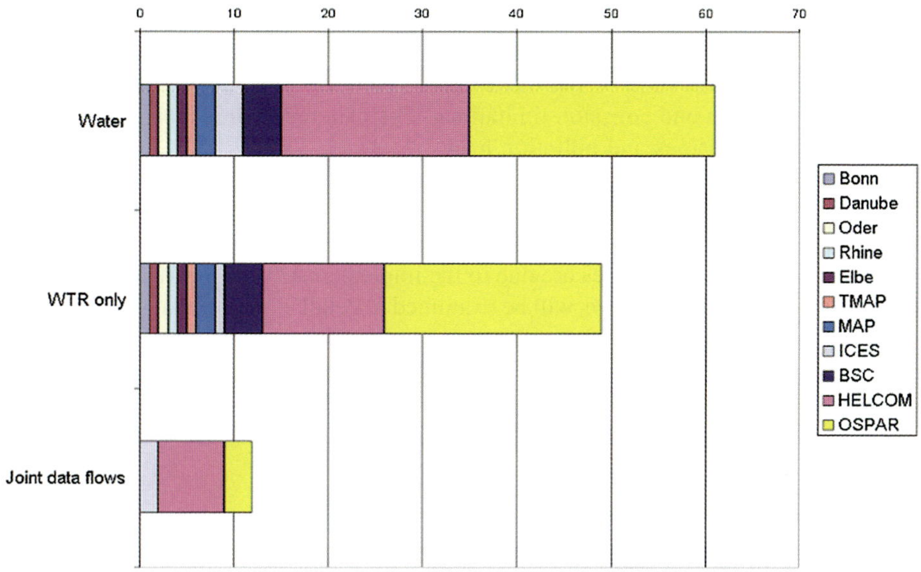

Fig. 3.2. Water data reporting to other data centres.

they are in the nature biodiversity area (the other obvious candidate for stream lining) (Figure 3.2).

There are five data flows that only constitute of one obligation (in ROD). However, four out of those are river conventions, with established monitoring stations and State of the Environment (SoE) reporting along them. They do constitute data flows, although limited to the riparian countries.

The other potential data centres in the water area are the marine conventions for the marine areas around the EEA member countries (Baltic, North Atlantic, Mediterranean, and Black Sea).

For HELCOM and OSPAR, it is known that some of the dataflows are being managed at the International Council for the Exploration of the Sea (ICES). It was found that though some of these dataflows originating from the conventions, the data is sent directly to ICES (See Figure 3.2). The remaining dataflows (obligations) are distributed in the following way:

	SoE	Mix
HELCOM	6	14
OSPAR	6	20

The SoE dataflows indicate that the data must be sent to the convention secretariats. It may be possible that the conventions then subsequently forward the data to ICES. This is not known at the present point in time. The convention web sites[1] do reference some of the data, and as such they act as "data centres" as well as ICES.

The compliance and state of the environment dataflows are mainly concerned with measures for pollution and emission limitations. The data reported is then concerning the actual values for emissions and pollution load.

All in all it seems that all of the HELCOM and OSPAR dataflows identified are valid. Although the extent to which ICES manages all the data could be further investigated.

It has been anticipated from the ETC/Water that major changes/revisions will be seen in the next few years. These changes are due to the implementation of the Water Framework Directive (WFD). These dataflows will be examined in detail as part of that work.

The WFD sets milestones along the way of building a new information system, including proposals for revising legislation, milestones to provide access to the public (late 2006), provisions for getting the system distributed (late 2009) and finally to rely on services provided by member countries.

3.6.2 Review of Water Monitoring

It is a key task to look at the ways water information is made available, and how it is aggregated, compared, and analysed.

Various activities are ongoing to (e.g. reviewing river catchments approaches) coming up with an improved representation of monitoring stations and other measures. This work is guided by suggestions from the WFD.

3.7 FROM A REPORTING TO AN INFORMATION SYSTEM

For some years, institutions involved into water reporting, foremost DG Environment, the Joint Research Centre, EEA and Eurostat, have been discussing the way forward in this area. Especially in 2004 and 2005, much progress was made to generate a common understanding about the role of the different partners, and on how such an information system looks like.

An information system consists of different components to provide different types of functions levels

- Reporting level
- Visualisation level
- Data retrieval or services level.

The plan is to realise such elements during the phases 1 and 2 until about 2009. From 2010 onwards, fully distributed services shall emerge.

Some key principles and elements of the common understanding on what is being called the Water Information System for Europe (WISE) are:

- Planning shall follow the implementation path of the WFD.
- A multitude of players – beyond the abovementioned, so-called "Group of four" – is involved, so a distributed system is what needs to be built.

- Success also depends on the way the member states can be involved in this. Issues on properly aggregating regional and national data up to the European level are very challenging.
- The distributed system will need some common entry points for customers with different interests. One could also call them *services*, and they need to be properly defined.
- Current dataflows have to be analysed, adapted, and included under this umbrella of services.
- The work will benefit from INSPIRE recommendations on spatial data handling; in fact, since WISE is one of the first implementation areas, building this information system will benchmark INSPIRE ambitions as it will do for the ones around the Shared Information System (SIS).
- Further standardisation efforts are necessary (e.g. the role of distributed data dictionaries has to be explored). The more those dictionaries are distributed, the more they need to follow standard ways in describing the data in them. EEA has gained experiences with standards around ISO11179. More promotion has to be done on such and related standards to assure their further proliferation. State of the Environment water data has been systematically recorded in the Reportnet data dictionary and is used to support data exchange.
- When more and more experts from different communities share water-related data ("talk about water-related issues"), the need to harmonise their language, to explain the concepts they have in mind and use, is growing. Technically, this translates into a need to build and support glossary and thesauri, when the discussions get more advanced ontologies.

NOTES

[1] *http://www.helcom.fi/ http://www.ospar.org/*

Chapter 4

AGENT-BASED CONCEPTS FOR ENVIRONMENTAL DATA

J. Fowler and G. Pitts

4.1 INTRODUCTION

The World Wide Web (Berners-Lee, 1991) has transformed the culture and expectations of users in all disciplines. This has led to two challenges for information management. First, the huge amounts of data accessible via the Web make information discovery and retrieval a problem that is only superficially addressed by the development and use of Web search engines such as Google™. Second, the structured databases that were carefully crafted with a small, knowledgeable user community in mind are now potentially accessible and useful to many additional users who are not so familiar with database query languages, even though they are comfortable with the use of the Web. The result is a need for information systems that can uniformly deliver both structured and unstructured data; can access different structured databases using a single query language and logical structure and return results derived and merged from distributed sources; and can at the same time communicate with users employing software standards and behaviour metaphors that they are comfortable with.

Environmental information systems are no exception to this Cultural Revolution. Currently it is very difficult to share environmental data, since the information typically resides on geographically disparate and heterogeneous systems. Because they were designed with little expectation of interoperating or of widespread access, these systems often do not facilitate access by secondary users. This can frustrate attempts to fuse data from multiple sources in the interest of arriving at a comprehensive understanding of environmental conditions and actions.

The US Environmental Data Exchange Network (EDEN) Project was undertaken from 1997 to 2000 by the Microelectronics and Computer Technology Corporation (MCC) to address these issues in the environmental domain. The project was a framework for cooperative activities relating to international environmental security, including activities such as information exchange and technology demonstration.

The EDEN Project utilised an emerging information technology called InfoSleuth, which was developed by MCC. InfoSleuth agents served as "intelligent" environmental information services that provided uniform access to specific sets of geographically distributed environmental information databases through standard Internet browsers

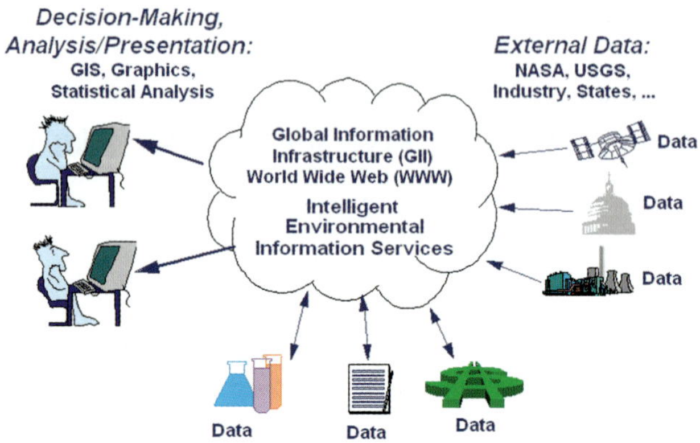

Fig. 4.1. Utilizing the WWW and the developing infrastructure to access distributed information sources.

(see Figure 4.1). To be effective, this needed to go well beyond a "keyword" model that did not comprehend context, as typically implemented in the Web search engines. It was also desirable to access, retrieve, and process the actual data in data resources, not simply return an address where the data may be found, as Web indices do when they return hypertext links using universal resource locators (URLs).This chapter describes the U.S. EDEN project, beginning with a more detailed motivation of the work and an overview of the InfoSleuth system. A discussion follows of the technology used by InfoSleuth and of a functional example to illustrate its use. The EDEN Project plan is described in more detail, followed by a description of benefits for the project participants.

4.2 PROJECT MOTIVATION

There were several factors that led to the creation of the EDEN project:

- U.S. federal statutes mandated that the involved US regulatory agencies, DoE, DoP and EPA, and departments cooperate with each other in sharing information about environmental issues. Each is intimately involved in environmental management and remediation, which requires access to objective and reliable information. In addition, the European Union and the United States share an interest in cooperating on remediation of contaminated sites at military bases in Europe.
- The reduced size of the U.S. governmental budgets necessitated the sharing of resources and information where possible. Where duplication of information and services exists, distributed access methods can reduce the need for that duplication. Where duplication is still appropriate, automated processes that compare the contents of different databases can point out inconsistencies and errors, ensuring that decisions can be made on the basis of the database with increased confidence in the validity of the contents. In addition, where interagency reporting can be simplified, money can be saved.

- The U.S. EDEN Project participants recognised their responsibility to facilitate the public's right to know. Increased ability to support public access under the Freedom of Information Act can substantially elevate the utility of the project.
- As a natural result of the fact that these systems have grown up in the hands of different developers, comparable data in these different systems is not necessarily stored in comparable data structures. For example, where one database may refer to U.S. states by two-letter abbreviations, another may refer to them by numeric codes, or by full name. This value-mapping problem is present whenever two or more information systems are integrated.

A number of benefits were anticipated from the collaboration of the government participants, which extended beyond the direct participation in the EDEN Project. The collaboration was expected to do the following:

- Provide an opportunity for interagency cooperation in developing the best available technology for overcoming major problems in accessing environmental data.
- Increase the efficiency of environmental information gathering, both by decreasing the amount of effort necessary to integrate and standardise data sources and by simplifying the task of an end user in querying the available data.
- Expand the environmental knowledge base without replication, enabling the sharing of information without incurring the cost of restructuring existing data systems or building new ones.
- Encourage standard environmental terms and definitions by promoting the development of a useful multilingual environmental thesaurus and then using it in the management of distributed data.

Utilising MCC's InfoSleuth technology, EDEN was intended to assist government agencies in integrating disparate databases, thereby allowing more effective access to and analysis of the data they contain.

4.3 INFOSLEUTH

Microelectronics and Computer Technology Corporation's InfoSleuth Project and Environmental Programs collaborated to provide the underlying technology for the EDEN pilot system. The InfoSleuth Project developed an "intelligent" agent technology to automate access to distributed information resources through Internet-capable browsers.

Database research in the past has focused on the relatively static environments of centralised and distributed enterprise databases. In these environments, information is centrally managed and data structures are fixed. Typically, the integration of concepts to specific sets of data is well known at the time a database schema is defined and data access can be optimised using pre-computed approaches. Federated database systems, although they support distribution of the resources across a network, do not depart from the centralised model of a static database schema.

At the other extreme, the Web presents a different paradigm. On the Web, there is a tremendous amount of textual information, spread over a vast geographic area. There is no centralised information management because anyone can publish information on the Web, in any form. Thus, there is minimal structure to the data. What structure there is may bear little relationship to the semantics. Therefore, there can be no static mapping of concepts to structured data sets, and querying is reduced to the use of search engines that locate relevant information based on full-text indices.

The InfoSleuth Project at MCC broadened the focus of database research to produce a model for combining the semantic benefits of structured database systems with the ease of publication and access of the Web. This change in fundamental requirements dictated a pragmatic approach to merging existing research in database technology with research in other computer disciplines. The result was an architecture that operated on heterogeneous information sources in an open, dynamic environment. Information requests to InfoSleuth are specified independently of the structure, location, or even existence of the requested information. A key to the success of this approach was the development for each application, a unifying "ontology" for the application domain. A domain ontology is a set of concepts, relationships, and meta-information that describe data at a high level and provide useful conceptual links between the elements of the data. This enabled the user to bridge the gap between different notions of data and the different schema of databases. InfoSleuth accepted requests specified at a high semantic level in terms of the global ontology, and flexibly matched them to the information resources that were available and relevant at the time the request was processed.

4.4 TECHNICAL SYSTEM DESCRIPTION

InfoSleuth was an extension of previous MCC work, namely the Carnot Project, which was successful in integrating heterogeneous information resources in a static environment. In this previous work, MCC developed semantic modelling techniques that enabled the integration of static information resources and pioneered the use of agents to provide interoperation among autonomous systems.

The InfoSleuth Project extended these capabilities into dynamically changing environments, where the identities of the resources to be used may be unknown at the time the system is designed. InfoSleuth, therefore, respects the autonomy of its resources and does not even depend on their presence. Information-gathering tasks are, therefore, defined generically, and their results are sensitive to the availability of resources. Consequently, InfoSleuth must provide flexible, extensible means to locate information during task execution, and must deal with incomplete information.

4.4.1 Technological Developments

To achieve this flexibility and openness, InfoSleuth strove to integrate the following technological developments in supporting mediated interoperation of data and services over information networks:

- *Agent technology*: Specialised agents that represent the users, the information resources, and the system itself cooperate to address the users' information-processing requirements, allowing for easy dynamic reconfiguration of system capabilities. For instance, adding a new information source should involve merely adding a new agent and advertising its capabilities. The use of agent technology permits a high degree of decentralisation of capabilities, which is the key to system scalability and extensibility.

- *Domain models (ontologies)*: Ontologies give a concise, uniform, and declarative description of semantic information, independent of the underlying representation of the conceptual models of information bases. Domain models widen the accessibility of information by allowing multiple ontologies belonging to diverse user groups.
- *Information brokering*: Broker agents match information needs, specified in terms of some ontology, with currently available resources. Retrieval and update requests can then be properly routed to the relevant resources.
- *Internet computing*: Java programs and applets are used extensively to provide users and administrators with system-independent user interfaces, and to enable ubiquitous agents that can be deployed at any source of information regardless of its location or platform.

4.4.2 Agents

The InfoSleuth system employed a number of intelligent software agents to perform its tasks, concealed from the user by a dedicated user agent. These agents operated independently in a distributed fashion and may be located anywhere over the network in the form of a Java program. Each agent provided a critical capability in the overall system, as described below.

- *User agent* – constituted the user's intelligent gateway into InfoSleuth. It used knowledge of the system's common domain models (ontologies) to assist the user in formulating queries and in displaying their results.
- *Ontology agent* – provided an overall knowledge of ontologies and answered queries about ontologies. This permitted users to explore the terminology of the domain and learn to phrase their queries in a way that will allow the application to provide useful results.
- *Matchmaker agent* – received and stored advertisements of the capabilities of all InfoSleuth agents. Based upon this information, it responded to queries from agents as to where to route their specific requests.
- *Resource agent* – provided a mapping from the global ontology to the database schema and language native to its resource, and executed the requests specific to that resource, including subscription queries and notification requests. Resource agents exist not only for structured databases, but also for unstructured data sources that serve text or images.
- *Task execution agent* – coordinated the execution of high-level information-gathering subtasks necessary to fulfil queries and other information management tasks, such as the control of workflow processes.
- *Multi-resource query agent* – used information supplied by the matchmaker agent to identify the resources likely to have the requested information, decomposed the query into pieces appropriate to individual resource agents, delivered these sub-queries to the resource agents, and then retrieved and reassembled the results.

The matchmaker agent's knowledge base of agent advertisements might be described as similar to the library's card catalogue and holdings locator list. However, this only describes the matchmaker's syntactic reasoning abilities. In addition, the matchmaker's constraint-based reasoning engine allows it to compare the metadata given to it as agents advertise their capabilities with the conceptual constraints imposed by a particular query.

To describe this in more detail, when a query is delivered to the multi-resource query agent for handling, the query agent requests a list of resource agents that may be able to answer individual fragments of the query. The names of those resource agents that have advertised knowledge of the entities and relationships that compose the query are returned. In addition, however, if an individual resource agent has advertised to the matchmaker that its data is constrained to a certain range (as an example, "I have information about NPL sites, but only about those in Texas"), the matchmaker will not return the name of that resource agent when the query is confined to the state of Pennsylvania. Judicious use of constraint-based advertising and reasoning can improve the scalability of the system by reducing bandwidth and computation for queries that, while being syntactically correct and semantically appropriate, are logically certain to return no results.

The agents in EDEN communicated with each other by means of the domain ontology for hazardous waste. For a given conceptual domain (such as chemical names and codes), this ontology specified the particular value domain (Chemical Abstracts Services number, for example) to be used in inter-agent communications. Both the user agent and the resource agent, however, provided interfaces to other systems, and might require value mapping to translate the terms of the user's value domain (chemical name) or that of a particular database (CAS number with embedded dashes, perhaps) into and out of the canonical value domain of the ontology. In EDEN, this value mapping was performed by a call to a value-mapping agent. Because the resource and user agents dealt with this translation issue, the multi-resource query agent was able to collate and join results knowing that the results were already consistent with each other.

4.5 PROJECT DESCRIPTION

The broad collaborative effort among the U.S. EDEN Project participants encompassed two major areas. One area was describing the "content" of environmental information. What is it? What does it mean? What is its quality and utility? Why was it created and how? What specific data is desired from the data sources? The second area was developing the means for sharing this information without incurring the financial and technical burdens of redesigning a federate database system or maintaining redundant databases. This was the focus of the U.S. EDEN Project.

The U.S EDEN Project participants chose to develop the pilot demonstration in the area of hazardous waste remediation. Each government agency contributed one or more database portions whose contents were made accessible through InfoSleuth. The databases that were identified for inclusion in the pilot included the following: CERCLIS (Coordinated Emergency Response Cleanup Liability Information System), ITT (Innovative Treatment Technologies), and HazDat (Hazards Data), from the EPA; IRDMIS (Installation Restoration Data Management Information System) and ERPIMS (Environmental Resource Program Information Management System) from the DOD; the DOE's EM-50 technology database; and the EEA's Basel Convention database. CERCLIS provided summary data concerning site characterisation and contamination of sites on the Superfund National Priorities List (NPL). ITT contained information relating to technologies for remediation of chemical contaminants, chiefly emphasising sites on the NPL. HazDat related chemicals to their health effects on humans. IRDMIS and ERPIMS provided site-specific data on contaminated sites at military installations managed by the Army and Air Force respectively. EM-50 was a primarily textual database of remediation technologies maintained by the DOE Environmental Management office. The Basel Convention database related to data describing trans-boundary shipment of hazardous waste in accordance with the Basel Convention treaty on that subject.

The EDEN Project also leveraged the work being done in two other important environmental information projects. One was the EPA's Environmental Data Registry (EDR) for describing data elements, which serves as an important foundation for addressing the value-mapping problem between data sources and the global ontology. This work related to proposed international data and metadata standards, such as ANSI X3L8 and ISO SC14. The other project, GEMET (the General European Multilingual Environmental Thesaurus), developed by the EEA, provides structured knowledge that makes possible translation of queries between European languages.

4.5.1 EDEN Development Plan

The development plan for the EDEN pilot involved the following steps:

- Creating a conceptual model that will become the application's domain ontology. To the extent possible, the ontology was to be constructed using terms that are applicable to the domain of waste remediation from the GEMET thesaurus. A graphic depiction of the ontological model that supports a set of queries chosen to show off the capabilities of the pilot is found in Figure 4.2. This figure shows the primary entities of the ontology linked by a collection of relationships that provide the best abstraction of the data to be found in the databases chosen for the pilot.
- Developing mappings between each of the identified databases and the domain ontology, and then configuring a set of resource agents, each of which uses the appropriate mapping to translate between its database resource and the common vocabulary provided by the domain ontology.
- Taking advantage of the EDR to assist in resolving issues arising in value mapping between the different ways different database designers may have used for storing data values that express the same concept (such as conversion between English and metric measures or reconciling different ways of identifying chemical contaminants).

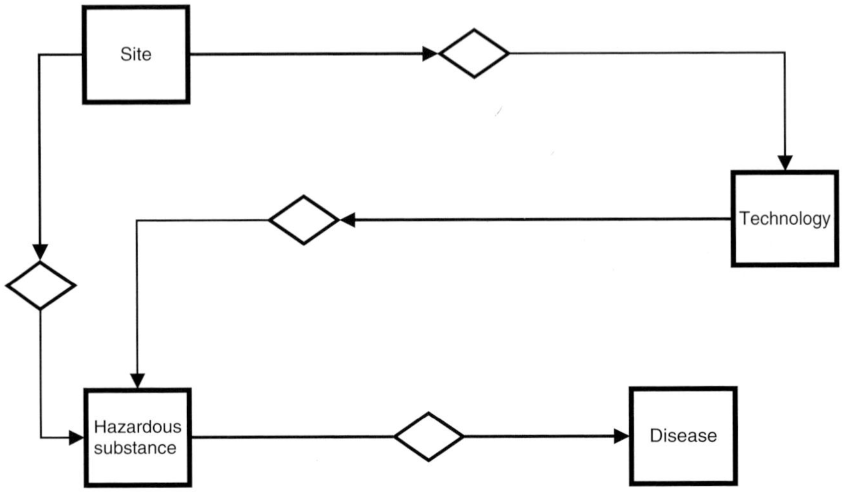

Fig. 4.2. The ontological model that supports a set of queries for the selected databases.

- Developing a flexible yet simple query tool that allows a user to pose queries over the domain ontology and retrieve answers whose appearance may be customised for an individual or group. The user interface was constructed using Java, which made it portable across numerous operating systems and graphic user interface environments. By using Java, it is possible for an end user to issue queries to the system with no more hardware or software than is necessary to support a Java-capable Web browser. Figure 4.3 depicts a user interface that has been configured to support a particular sequence of queries relating to identifying remediation technologies associated with particular sites. The results displayed were retrieved from a demonstration system containing data from CERCLIS and IRDMIS.

EDEN utilised intelligent InfoSleuth software agents and Java applets to access and retrieve information from disparate data sources. The technology was designed to support a dynamic environment in which databases can be added or removed without affecting the basic behaviour of the system. Thus, the project could be developed from a small initial group of databases, with additional databases being added to the project. The resulting technology demonstration was useful in providing access to distributed environmental data resources via the Internet, as well in guiding information systems developments such as EDEN-IW.

EDEN demonstrated a simplification of the exchange and sharing of information among organisations through common vocabulary and generalised query ability. By establishing the common vocabulary, widely differing information resources can be "mapped" and readily accessed by a sophisticated system of software agents that involve in advertising, matchmaking, and the exchange of the data requested by the user. In this way, EDEN

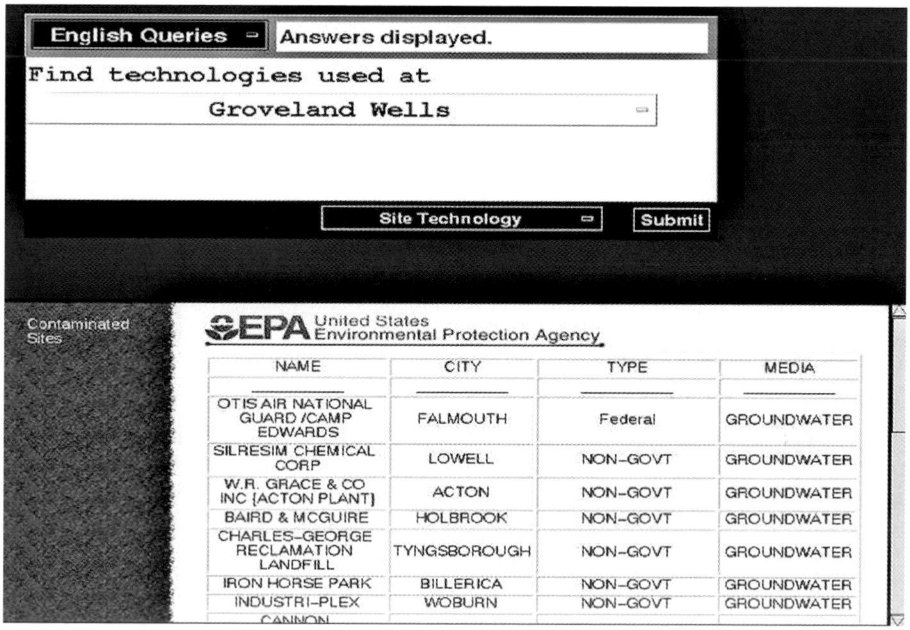

Fig. 4.3. A sample user interface for identifying remediation technologies.

provided uniform access to existing information resources without imposing requirements for restructuring or incurring the significant cost of conventional database integration.

The goal for the user interface to EDEN was to provide a tool by which templates could be created that permit query formulation based on sentence-based queries that contain embedded pick-lists for the parameters of the query. These queries were translated by means of the template into Structured Query Language (SQL), the lingua franca of InfoSleuth. The use of these templates allowed environmental domain experts and other users to serve as application developers, by constructing user interfaces without the need to write user interface code.

4.6 CONCLUSION

Several unresolved issues with the EDEN prototype exist that served to inspire further work in EDEN-IW. The ability to communicate how the system works to persons who are not experts in the fields of distributed information systems and knowledge management suffered because of the inherent complexity of the system, both from a logical point of view and with respect to the impact that its physical distribution had on its performance.

Demonstrating the efficacy of EDEN's infrastructure was a challenge because, to the user, a well-performing system appears to have no appreciable difference from a monolithic system. When the user issues a simple request, using an interface similar to what they might use on a database wholly contained on his or her own machine, there is a pause and then a set of results is returned. The user is shielded from the complex message interchange that takes place between the various agents once the query has left his or her screen and entered the cloud of InfoSleuth agent communications (as previously depicted in Figure 4.1). However, viewers of the system in action perceived a significant need for a visual tool that showed geographic and logical relationships between the agents that communicated to resolve a query.

The EDEN system demonstrated the feasibility of investigating distributed environmental data using an agent system that communicates concepts through a global domain ontology. At the same time, EDEN fell short of providing a truly useful understanding of the breadth and depth of the domain ontology. The meanings of concepts and the relationships between them were not easily extracted from the system; in addition, the precision of the ontology in expressing subtleties relating to environmental determinants needed significant amplification to address the real needs of scientists and administrators. These issues demanded significant improvement, not only in the expressiveness of the domain ontology and the creation of modules to reason over it, but also in the ability of the user interface to render these subtleties.

These issues aside, by demonstrating the plausible prototype of a common instrument to improve the understanding of environmental problems, the U.S. EDEN Project stimulated collaborative environmental security activities between government agencies in the U.S. and Europe, as well as facilitating information sharing between U.S. and European environmental professionals.

The potential for an effective means of managing and sharing environmental information is enormous. The Web provides clues as to what distributed information systems of the future can offer, but by no means can it be said that the Web itself is a solution to cooperative information management. Great strides remain to be taken in support of collaboration and intelligent information retrieval.

The U.S. EDEN Project was an ambitious effort to demonstrate the use of an agent-based system to improve the ability of a critical set of government agencies from both sides of the Atlantic to collaborate in the management of environmental data. As an application of the InfoSleuth intelligent agent technology developed at MCC, it represented a significant step forward in demonstrating the potential to organise, access, and analyse environmental information. The collaboration between the participants is stimulating the development and adoption of appropriate data standards and methods for describing data elements that reached well beyond the U.S. EDEN Project, including stimulating the creation of a follow-on effort, the EDEN-IW Project.

ACKNOWLEDGEMENTS

InfoSleuth was an MCC consortial research project sponsored by General Dynamics Information Systems (formerly Computing Devices International), Motorola, NCR

Corporation, Raytheon Company, and TRW Inc. The EDEN project was supported by the National Institute of Standards and Technology contract 50SBNB6C9076.

Much of this chapter was adapted from *Information Systems and the Environment*, from the U.S. National Academy of Engineering, 2001. Descriptions of the InfoSleuth technology were adapted from prior publications listed in the References. The authors wish to thank Mike Minock, Malcolm Taylor, Marian Nodine, Qing Jia, Brad Perry, Nancy Perry, and Vipul Kashyap at MCC for their contributions to this work.

Chapter 5

REVIEW OF MODELS AND TECHNOLOGIES FOR DATABASE INTEGRATION

S. Poslad, M. Stjernholm, L. Zuo and X. Huang

5.1 INTRODUCTION

Currently, despite the onslaught of the Web, much information, such as transaction-based environment data, is collected periodically and exists in a form such as a database. This cannot be accessed directly, interlinked or searched using standard Web tools such as search engines, although specialised tools could be developed to support Web-based database access and integration. Databases are very sophisticated and mature data storage and retrieval systems that unlike Web data are explicitly designed and are able to support high data quality. Often as databases are developed and managed autonomously, they are heterogeneous. It can be useful to combine data from multiple databases, e.g. to determine if environment quality indicators vary across different geographical regions (databases).

There is a range of potential models and technologies that could support both database query transparency and database (system) integration of multiple distributed heterogeneous databases. By database query transparency, it is meant that users need not be concerned with:

- Differences in national languages used to retrieve and store the data;
- The location and access details of the database;
- The addresses of data within the database; the peculiarities of the relational database management system (RDBMS), e.g. vendors often use slightly different Structured Query Language (SQL) interfaces, user and management interfaces;
- The schema (a structured model of data expressed in a standard representation) used to organise the data for storage and retrieval at the SQL level.

Query transparency and database integration require a data model that can reach across multiple heterogeneous databases. Typically, this is a metadata model – metadata is data that describes and indexes data to improve storage and retrieval performance, and categorises the stored data. The metadata model also requires associated middleware services that can be used to manage such data. There are several approaches to integrating databases. We consider the most significant metadata-based models to be: SQL-based RDBMS schema, XML-based schema and Ontology or Semantic Web-based schema. Middleware services to support these can be based upon data warehouses, distributed databases, Web Services, Grid and Multi-Agent Systems (MAS) architectures. The advantages and disadvantages of each are pointed out below.

5.2 RDBMS, SQL AND DATA WAREHOUSES

The SQL, standardised by ANSI (American National Standards Institute), is the current standard language for querying data from all major RDBMS. SQL is a very powerful database user interface and API (Application Programmers Interface), because it gives the user the flexibility to form new queries at run-time, without necessarily requiring database applications to be reprogrammed. It is essentially a data verification technique; it uses predetermined queries and verifies the query in terms of whether results that are returned answer that query or not. SQL is underpinned by a formal, mathematical model, and it uses simple arithmetic operations such as LIKE or EQUALS, (relational) algebra operators such as JOIN and PROJECT, and logical operations such as NOT, AND and OR.

SQL, whilst underpinned with a (relational algebra) mathematical model, contains redundancy in that different query syntaxes can be used to form the same query. One example is the SQL WHERE clause compared to the different variants of the JOIN operator. Although SQL has been standardised by ANSI, there are different versions of SQL and different database vendors selectively support different parts of that standard. It can be argued that very few RDBMS support the full standard SQL model as most, e.g., do not implement the concept of a data domain. Although SQL defines data syntax when creating the structures for storing data in a database, it does not define data semantics, i.e. the meaning of the data within an application context.

In some cases users may not know how to form queries to retrieve specific data, or which tables contain the relevant data, or even which databases contain the relevant data. Users may need to do a more general search to select data, rather than to use prior knowledge to make specific queries. Therefore searches are more efficient if they are made on metadata rather than the data itself. The SQL standard supports metadata to describe the data syntax. SQL metadata can also be stored as tables in the database. SQL queries can then be used to query the metadata tables, in the same way that they can be used to query the data tables. However, use of SQL for searching databases rather than querying is restricted because of the lack of a common specification for metadata descriptions and semantics; the lack of provision of metadata in individual database instances; and the lack of a standard namespace to locate tables within a database and to locate tables across multiple databases. In addition, metadata that describes the format of the data tables is often defined by the database applications that create the data, but other query applications often cannot access this.

A further complication is that often the design of a sub-domain of the world, such as inland water, is more accurately modelled in terms of an object-oriented hierarchy, but the relational data schema that define the database structure are organised as a single tier of flat table structures that have no hierarchy (see Figure 5.1). This leads to mapping functions being required to translate from a world model to the flat table data storage model.

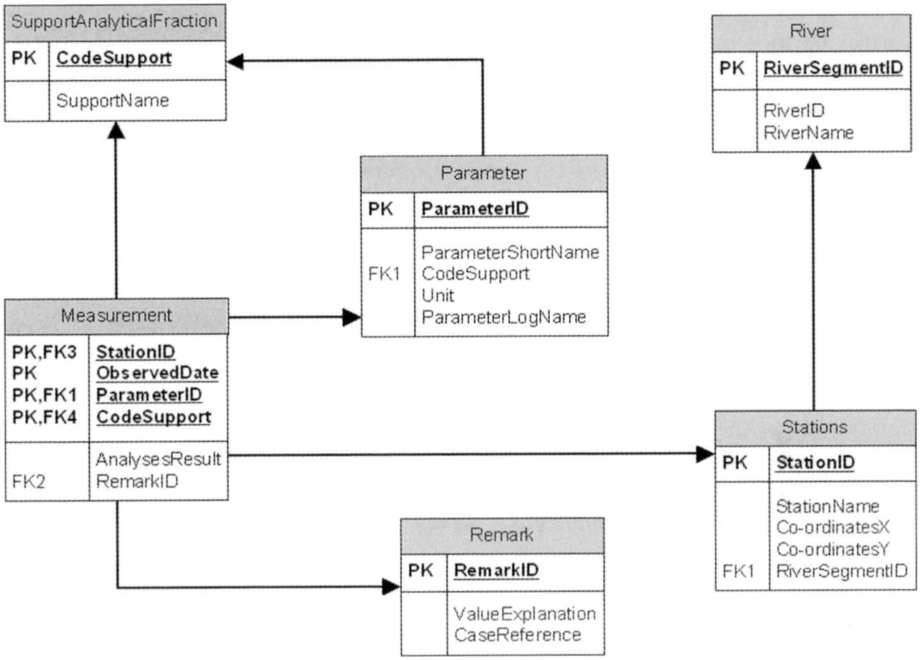

Fig. 5.1. The database schema of an inland water database from the International Office for Water (IOW).

A key feature of RDBMS is that unlike Web documents, data is created and maintained to support high data integrity (by protecting data existence, isolating transactions and supporting data backup and recovery), data quality (so that the data is accurate, complete and current) and ensuring confidentiality. To maintain data quality, several techniques are used such as integrity constraints, data validation, concurrency update control and update authorisation. RDBMS also provide tools to ease data creation and data management (by maintaining data quality), but there is little support for database integration and query transparency across multiple heterogeneous databases.

5.2.1 Database Integration Overview

A major challenge for integrating multiple databases is the semantic heterogeneity among different databases. This is difficult to handle due to the autonomy and semantic heterogeneity of the component databases. Methods for schema analysis, transformation, integration and mediation have been investigated in the database community, in order to produce tools to handle this problem. However, schema level analysis is often quite a manual process and may sometimes be too general to solve the problem. Data level analysis, i.e. the analysis of database contents, is needed, e.g. to support the collecting and harmonisation of data where the same concepts are modelled differently in different databases.

There are two basic types of methods for integrating heterogeneous data: either their individual data models are merged to a common or global view or the individual databases are aligned to each other without using a common form. Usually the databases are merged using a common form, as often fewer mappings exist this way. However, a difficulty with using the common form is how to deal with data that is not common to all individual databases.

5.3 THE GLOBAL VIEW DATABASE INTEGRATION MODEL

Figure 5.2 shows the layout of a general Global View Database Integration Model. The bottom layer consists of a resource management layer of database sources defined by local (database) schema. Each of these gets mapped to the same external or global schema that can be used by multiple applications in the middle layer, and can be presented in multiple viewpoints in the top layer.

5.3.1 Data Warehouse

The data warehouse (see Figure 5.3) is an integrated, time-variant and non-volatile set of data that supports decision-making (Immon, 1996). Data integration is a complex process and involves extracting a data subset for export, from sources such as databases, transforming it into a standardised format and cleaning up the data to remove inconsistencies, redundancies and errors. The data warehouse is periodically updated with a bulk upload, rather than when every single transaction occurs. Data warehouses are generally used for

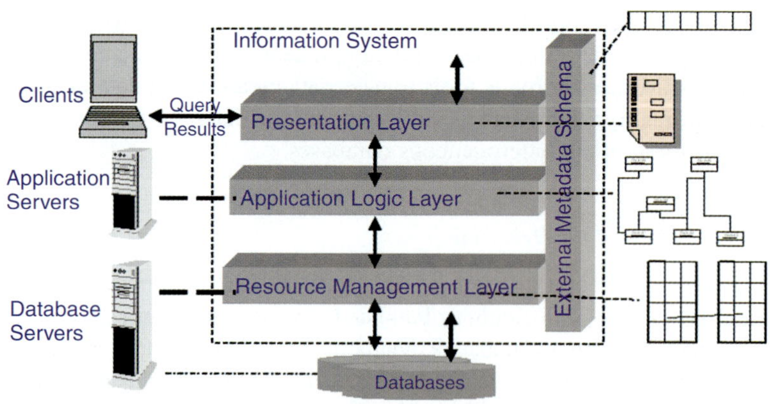

Fig. 5.2. Schema architecture of a layered heterogeneous database system.

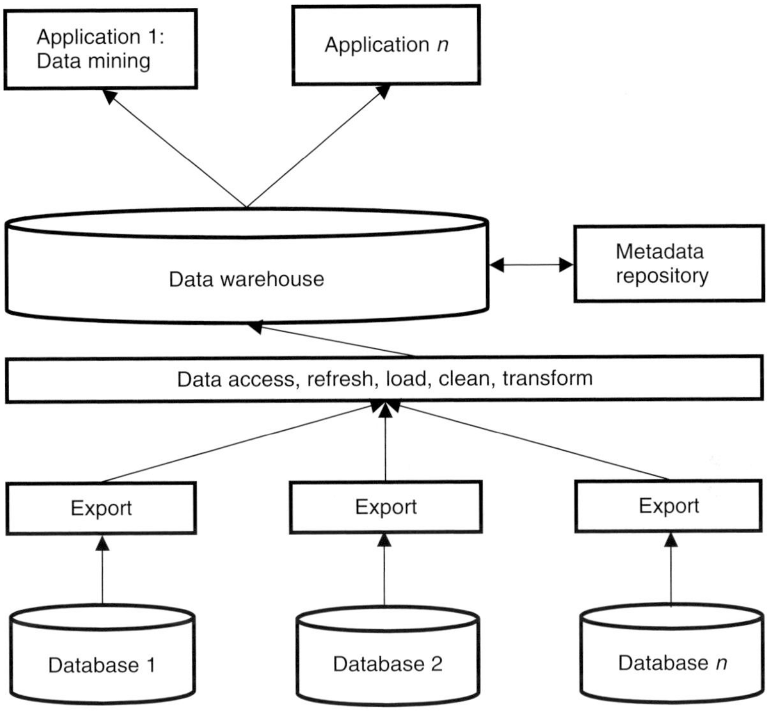

Fig. 5.3. A multi-layer information architecture.

online data analysis such as data mining, rather than for answering data queries. Generally, all the different data sources fed into a data warehouse are owned and managed by a single enterprise. This is where it differs from a set of environment databases (such as the ones used in the EDEN-IW project), which are owned by autonomous national and scientific data collection entities. It would require lengthy negotiations to get the different EDEN-IW database owners to input their data into a data warehouse or to delegate rights to others to manage copies of their data. In addition, resources to create, operate and maintain the data warehouse are needed.

5.4 DISTRIBUTED DATABASES

A distributed database system enables multiple databases to exist at multiple locations, but to be queried as if they were centrally located, and without the need to export partial copies of the data to a common data (warehouse) store. Distributed databases can easily join multiple distributed data that is fragmented and replicated across multiple databases. A major challenge for the fragmentation and rejoining to work is that the data fragments need either to have the same data schema (horizontal fragmentation), or to be subsets of a consistent higher level data schema (vertical fragmentation). Hence this is not usable if data schemas in different databases are not compatible in this way. Distributed databases are supported as extensions of existing RDBMS.

5.4.1 Database Application Programming Interface (API)

The need to access single or multiple database data in different applications has driven the standardisation of a generic API to wrap SQL so that it can be used by applications – wrapping involves creating and associating a programming language call with each SQL statement such as select or create. In 1996, a standard database API was adopted as an international standard called X/Open CLI (Call Level Interface) (Watson, 2004). An extended implementation of CLI was developed by Microsoft. It is called Open Data Base Connectivity (ODBC), and has become the standard cross-platform API to access any SQL database that supports an ODBC (driver). Java database connectivity (JDBC), a Java portable version of ODBC, has also been developed to allow Java applications to access databases. A powerful feature of JDBC is that calls exist so that it is possible also to access database metadata. In this way it is possible to obtain a list of tables, which are available in the database, and also to obtain information about the variables in these tables. Note that JDBC gives metadata of the structure of the tables – but not of the semantic of the table data. It should also be noted that JDBC may not always be available to access RDBMS. The use of JDBC and ODBC makes it possible, by higher-level service abstractions such as Grid and software agents, to access and process database data.

5.5 SQL SUMMARY

Relational type data models and associated middleware services can be considered as a low-level approach to the integration of heterogeneous IW databases. However, while SQL specifies a standard syntax for queries, it lacks a standardised syntax for metadata and for the results of the queries. Related middleware – such as data warehouses – requires data management and access rights to be delegated to third parties, or requires homogeneous database schema. Middleware also lacks a standard name and directory service to improve searching for databases and tables within the databases.

5.6 XML, WEB SERVICES AND THE GRID

5.6.1 XML

eXtensible Markup Language (XML) is a mark-up language, standardised by the W3C standards body that supports definition, transmission, validation and interpretation of data (XML-W3C, 2005). XML is one of the components required to exchange information in a universal format, but is not a complete solution to integrating heterogeneous databases.

Extensible Markup Language is a specification for computer-readable documents. Mark-up means that certain sequences of characters in the document contain information indicating the role of the document's content. The mark-up is user-defined and describes the document's data layout and logical structure and makes the information somewhat self-describing. It takes the form of words between brackets called tags, e.g. < name > or < h1 >. In this respect, XML looks very much like the well-known language, HTML.

The main mark-up entities in XML are elements and these elements are normally organised into nested hierarchies. The elements normally consist of an opening tag and a closing tag, e.g. < Determinand > and < /Determinand >. Elements might contain other elements or text. If an element has no content, it can be abbreviated as < Determinand/ >. Elements should be properly nested: a child element's opening and closing tags must be within its parent's opening and closing tags. Every XML document must have exactly one root element. Elements can carry attributes with values, encoded as additional "word = value" pairs inside an element tag. Thus, XML seems to be a powerful data model for the support of a standardised syntax for metadata, and for the results of the queries.

Extensible Markup Language does not imply a specific interpretation of the data. On account of the tag's names, the meaning of a piece of XML seems obvious to human users, but it is not formally specified. The only legitimate interpretation is that an XML code contains named entities with sub-entities and values. Thus, every XML document forms an ordered, labelled tree. This generality is both the strength and the weakness of XML. All kinds of data structures can be encoded in an unambiguous syntax, but XML does not specify the use and semantics of the data. The parties that use XML for their data exchange must therefore beforehand agree on the vocabulary, its use and its meaning.

In an additional XML schema document that can be associated with each XML document, a data model is defined. This data model describes the structure, i.e. which tags are required or allowed, where in the hierarchy they may or must be used and the tags that legally exist between the tags. Data that complies with the model is said to be "valid" and data that does not comply is "invalid". There are two mechanisms for creating data schema models: one is called a "Document Type Definition" or DTD and the other is a formal "Recommendation" released by the XML Schema Working Group of the World Wide Web Consortium (W3C) in 2001. Both DTDs and XML Schemas support the validation of the structure of an instance. Thus, both DTDs and schemas specify valid elements, their content models, valid attributes, valid attribute types and default attribute values. Schemas have the advantage over DTDs in that they can specify more constraints against which the XML data can be checked. In addition, an XML Schema is itself represented as XML whereas a DTD is presented using non-XML syntax. However, in most cases, it may not be necessary or even desirable to validate each XML message – the main processing may be just to parse and extract features from XML messages without the time-consuming validation.

At the application level, software APIs such as an XML DOM (Document Object Model) API can be used to parse XML data structures, whereas APIs such as JDBC can be used to parse database data structures and then map the equivalent data structures between the XML and the database models. A fundamental issue is how to mediate and map between the stored relational database data and the equivalent XML structure and vice versa, as there currently exists no standardised way to do this. There are two main ways of doing this – template-driven and model-driven mappings. In a template-driven mapping, there is no predefined mapping between XML document data structures and database structures, instead SQL commands are directly embedded as user-defined SQL elements in an XML document. Currently template-driven mappings are used for transferring data from a relational database to an XML document but not vice versa. In a model-driven mapping, a database table (or set of tables) can be modelled as an XML document of nested table row and column elements – or as some other XML tree. These XML elements generally correspond to table objects and attributes and adhere to an associated XML-schema corresponding to the database table data. Some sources estimate that more than 75% of current Web pages are generated from database data (Watson, 2004).

5.6.2 Web Services

There is a wealth of middleware (or Web service models) and specifications, proposed within the W3C standards consortium, to use and manage XML documents or data. These include: Simple Object Access Protocol (SOAP), a specification for XML-based message exchange, Web Service Description Language (WSDL), directory services based upon Universal Description, Discovery and Integration (UDDI) and Business Process Execution Languages (BPEL). Both open-source and commercial implementations of Web services are available. There is no comprehensive way to maintain the integrity of XML documents; limited integrity checks can be made by validating XML documents against any defined XML schema.

5.6.3 The Grid

Data Grids (Foster et al., 2001) are emerging as an important middleware model for managing data, and in a range of scientific and engineering disciplines that require computationally intensive analysis of large quantities of subject-specific data. The term “Grid” refers to technologies and infrastructure that enable coordinated resource sharing and problem-solving in dynamic, multi-institutional virtual organisations. This sharing relates primarily to direct access to computers, software, data, networks, storage and other resources, as is required by a range of collaborative computational problem-solving and resource-brokering strategies emerging in industry, science and engineering.

A data Grid system consists of a set of basic Grid protocols used for data movement, name resolution, authentication, authorisation, resource discovery, resource management and the like. A data Grid provides transparency in how data-handling and processing capabilities are integrated to deliver data products to end-user applications. This is done so that requests for such products are easily mapped into computation and/or data retrieval at multiple locations. The focus of the Grid software community is defining APIs at the Grid level to access databases. More recently the Grid community have based their architecture upon XML Web-service models to access and process data. For database integrators to take advantage of data Grids they need to install and operate complex Grid software such as that based upon the Globus toolkit.

5.6.4 XML Summary

Extensible Markup Language is a powerful model or schema, which provides the syntax to encode database queries and results data, for exchange. XML schema must be mapped and synchronised to the database schema. There is also a wealth of XML-based middleware services to support this. XML alone is insufficient to model the database information and the metadata used to search and integrate heterogeneous IW databases. Agreeing on a common syntax for structured data exchange, it can be argued, is the easy part. Agreeing on a common domain model of terms and their relationships is the hard part and hence multiple XML specifications can often exist for a given application domain. XML itself supports linearised hierarchical data structures, but because of its simplicity it lacks the expressivity to support inferencing, to explore and match data structures to support interoperability.

There are questions about the scalability of schema-based and syntactic approaches because of the number of possible heterogeneous schema and the difficulty in normalising numerous syntactical mappings between heterogeneous database schemas. As a result interoperability, based upon models of the semantics of the underlying databases, has been proposed as being more scalable (Kashyap and Sheth, 1994).

XML-based extensions, such as RDF and DAML + OIL (see below), have been developed that support more expressive metadata and richer inferencing. These also lack explicit communication protocols – most XML data exchanges use an implicit simple message template that includes both the request and the reply in the same message. Richer interaction patterns and communication protocols are needed to adaptively match user requests to service capabilities, to support service push as well as service pull and to support multi-party interactions and negotiations.

5.7 THE SEMANTIC WEB

5.7.1 Semantic Web

The word "Semantic" implies meaning, or is relating to the study of meaning and changes of meaning. For the Semantic Web, the term "semantic" indicates that the meaning of data on the Web can be discovered – not just by people, but also by computers. In contrast, although software can retrieve and search for relevant meaning, most of the meaning on the Web today is inferred by people who read Web pages and the labels of hyperlinks, and by other people who write specialised software to work with Web documents. The vision of the Semantic Web is to make data located anywhere on the Web accessible and understandable, both to people and to machines. The vision of the Semantic Web is much bigger than the building of the original World Wide Web. Semantic Web is a social force too. Companies must cooperate where they might normally compete.

Defining an RDF vocabulary: RDF Schema (RDF-S)
Basically, RDF Schema is a simple type system for RDF. It provides a mechanism to define domain-specific properties and classes of resources to which you can apply those properties.

The basic modelling primitives in RDF Schema are class definitions and *subclassof* statements (which together allow the definition of class hierarchies), property definitions and *subproperty-of* statements (to build property hierarchies), domain and range statements (to restrict the possible combinations of properties and classes) and type statements (to declare a resource as an instance of a specific class). With these primitives, a schema for a specific domain can be built.

The RDF Schema is a quite simple language compared to full-fledged knowledge representation languages (Klein, 2001). Also, it still does not provide exact semantics. However, this omission is partly intentional as the W3C foresees and advocates further extensions to RDF Schema. Because the RDF Schema specification is also a kind of metadata, RDF can be used to encode it.

5.7.3 Ontology Data Models

According to Gruber (1993), "An Ontology is a specification of a conceptualisation". Ontologies can be used to represent knowledge about domain concepts, and the relationships that are assumed to hold between them. It can describe the kinds of things that can be talked about in a context or domain. An ontology provides the means to classify things – to define a taxonomy. And it provides the means to define the properties and relationships that can be applied.

DAML + OIL (DARPA Agent Markup Language + Ontology Inference Layer) is an example of a specification for the Ontology component or layer in the layer-cake model (Bechhofer et al., 2001). DAML + OIL is well designed and naturally intuitive to (human) users. It also provides adequate expressive power, as it supports machine understanding and reasoning. It has a formal semantics (equally important) and extends existing Web standards. DAML + OIL is built on top of RDF(S). It provides basic ontological primitives to represent classes, relations (properties) and class (and property) hierarchies.

The aim of the DAML + OIL program is to achieve semantic interoperability between Web pages, databases, programs and sensors. The problem that DAML + OIL addresses is how to build a monolithic set of ontologies on agreed-upon domain models. The solution is to develop usable interoperability technologies similar to those that enable the Web to function. Towards this end, DAML + OIL will enable annotating information on the Web to make knowledge about the document machine-readable, so that software agents can interpret and reason with the meaning of Web information. DAML + OIL enriches and formalises metadata annotations (see Figure 5.4). DAML + OIL is the only part of the Semantic Web vision of the automation or facilitating of things that are currently difficult to do: locating content, collating and cross-relating content and drawing conclusions from information found in two or more separate sources. DAML + OIL's part is to serve as a markup language for network agents by providing a mechanism for advertising and

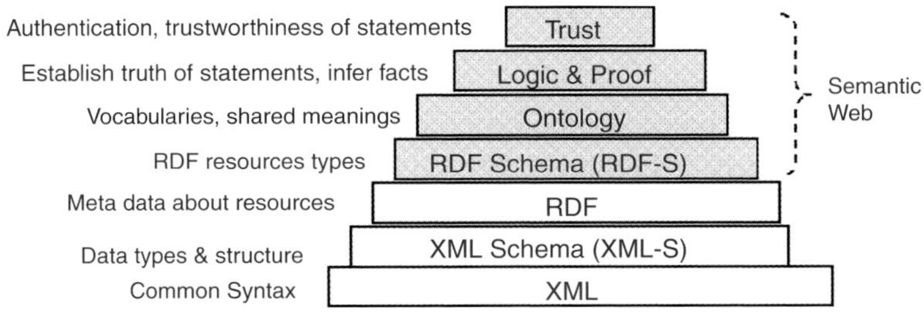

Fig. 5.4. Layered cake architectural model of the Semantic model.

reusing specifications. DAML + OIL's applications will be far-reaching, extending to both the military and the commercial markets. Its machine-to-machine language capabilities might be instrumental in realising the application-specific functionality, independent of human control. DAML + OIL will also enhance the efficiency and selectivity of search engines and other automated document processing tools.

In 2004, DAML + OIL was superseded by OWL Ontology Web Language as the W3C's candidate Ontology language (OWL-W3C, 2002). In terms of the conceptual model, this is broadly similar to that in DAML + OIL, however the logical inferencing capability to reason about the conceptual model is much improved in OWL.

5.7.4 Semantic Web Summary

RDF-based data models such as RDF-S, DAML + OIL and OWL are a powerful model or schema to provide the syntax to encode database queries and results data for exchange. The main focus of Semantic Web activity has been on developing expressing languages to represent the semantics and to support logical reasoning about the semantics. In contrast, the main focus of the SQL/RDBMS model is on managing data quality and supporting a high performance for data retrieval. A trade-off is often needed in practice, either a highly expressive semantic data model (that can be computationally very time-consuming to resolve queries) is chosen or a simpler metadata model that can lead to more robust data quality and better performance is chosen.

Use of a Semantic Web data model requires middleware services such as discovery services (McIlraith et al., 2001), protocols to parse, exchange and maintain semantic data models, and mediation to synchronise these to database models. These are not mature, yet are in widespread deployment at the current time.

5.8 AGENTS AND MULTI-AGENT SYSTEMS

An Ontology or Semantic data model also requires semantic data services and communication protocols to exchange and coordinate data, to import data models into applications and to parse it, to process semantic data, to support application-specific selection of data, to perform logical queries on data and to perform semantic mediation to map between different semantic data structures. We term a software entity that can perform the above-mentioned tasks a "software agent".

This is a specific form of the more general concept of an agent as a software abstraction that supports the properties of reactivity, proactivity, deliberation, social interaction and autonomy. Agents can autonomously monitor their own environment and take the action they deem appropriate. This characteristic of software agents makes them suitable for applications that can be decomposed into independent processes. They are capable of doing useful things without continuous direction by other entities. The autonomous ability coupled with an intelligent behaviour is further enhanced in a Multi-agent society. Agents can socialise using a rich set of standard interaction patterns. Communication enables the agents to coordinate their actions and behaviour, resulting in systems that are more coherent. Coordination involves cooperation and planning (centralised and distributed) (Weiss, 1999). Agent communication also involves knowledge exchange using a higher-level semantic model that is often based on ontologies. At a low level, agents use database APIs such as JDBC to access database data and then to embed this into more complex message structures at a higher level of abstraction of the agent (communication) level in order to exchange, coordinate and to reason about database data.

A multi-agent system is a good potential architecture for integrating heterogeneous databases in that agents are naturally distributed and autonomous; they can use rich explicit communication protocols to interoperate and they can naturally link to semantic models to help resolve interoperability problems. Multi-agent systems (MASs) have been and are the subject of a very active research community. We briefly mention a few representative examples of early proprietary agent systems:

- *InfoSleuth* (Nodine and Unruh, 1997) provides middleware in terms of an agent shell which includes a white-page directory service (library); an autonomous composite component, called the conversation layer, which provides routing, message-forwarding and basic dialog management; and a broker agent component. The agent system is implemented in a Prolog-like language called LDL++. The EDEN project, which was the forerunner for the EDEN-IW project, applied the InfoSleuth system.
- *JATLite* (Java Agent Template Lite) system (JATLite, 2005) provides Java middleware libraries, called layers, for basic communication service, a combined routing and message forwarding autonomous component or 'active library' and an agent communications library. The libraries can be substituted with alternatives. For example, the default basic communication library supports only TCP/IP transport and not UDP/IP and CORBA. However, it can be substituted by an alternative which supports these alternatives. Similarly, the agent communication library supports KQML by default but other alternatives can be supported.
- *KAoS* (Knowledgeable Agent-oriented System) system (Bradshaw et al., 1997) is designed to be independent of a particular communication service. KAoS agents are derived from a generic agent class (template-library type of middleware) which provides a basic communication mechanism. Several important agents may play a persistent role, but it is not clear whether this is implemented as middleware. Specialised middleware

agents carry out other generic services such as a matchmaker (yellow-pages) and domain manager (keeps track of ownership issues, white-page service); proxy and mediation agents act as external interfaces to the agent platform.

- *OAA* (Open Agent Architecture) system (Martin et al., 1998) middleware consists of an agent component called a facilitator which provides yellow-page directory, persistence and coordination services. OAA also provides an agent library, implemented in several languages such as Prolog, C, Java, Lisp, Visual Basic and Delphi, which is linked to each agent and offers the agent communication service via the facilitator. The proprietary communication language is called ICL and has a syntax like that of Prolog.

There are however interoperability problems, as none of these proprietary MASs are able to interoperate with each other. Furthermore, few (or none) of these proprietary MASs are open source. The highly interactive nature of multi-agent systems points to the need for consensus on agent interfaces in order to support interoperability between different agent systems, something which is necessary if MAS applications are to become pervasive. While it is challenging to develop MAS applications for a closed vertical architecture and market, it is even more challenging (and necessary) to develop MAS for horizontal MAS markets and open services.

One of the first standard models for agent communication was a result of the DARPA funded Knowledge Sharing Effort (KSE). It was developed at the University of Maryland around 1990 (Labrou, 1999). This model developed a three-layer model to define the pragmatics, syntax and semantics for sharing knowledge and hence supporting agent communication. The model was implemented using three different encodings: KQML, KIF and Ontolingua, respectively. In practice, the MAS model requires the addition of third-party agents called middle agents which facilitate communication by providing traffic in meta knowledge about the capabilities of service providers and the preferences of service user agents. This meta knowledge is supplied to (or extracted from) middle agents using a subset of communicative primitives such as Broker-one, Recommend-one and Recruit-one. As KQML developed, diversification occurred and different dialects and enhancements arose – there is no longer a single standard for KQML.

In 1996 a non-profit association, the Foundation for Intelligent Physical Agents (FIPA) (Poslad and Charlton, 2001), was formed to develop and ratify agent standards by common census (in contrast to the KSE program). The completion and adoption of such a standard is a prerequisite to the widespread commercialisation and successful exploitation of intelligent agent technology. In 2005, FIPA has become an IEEE standards body activity.

In some respects, FIPA can be regarded as a second-generation architecture for agents. FIPA shares some similarities with the KSE model, it has standardised the pragmatic layer encoding (ACL) and the [content] syntactical and semantic layer encoding. However, FIPA differs from KSE in terms of the functionality, which is standardised, and how this functionality is modelled. First, FIPA has standardised interaction protocols or common sequences of agent messages. Secondly, FIPA has specified an abstract model to allow agents to access non-agent services such as database access. Thirdly,

FIPA has specified several types of middle agent, such as a directory service, a name service and ontology service to support the exchange of semantic metadata. Fourthly, FIPA specifies rich semantic models for the use of its core speech act communication protocol. There are several open source and commercial implementations of the FIPA specifications.

5.8.1 Agents Summary

It is worth highlighting some of these MAS experiences in terms of challenges and solutions (Table 5.1).

Table 5.1. MAS challenges and solutions

MAS challenge	Solution
Proprietary heterogeneous MAS applications cannot interoperate.	Use MASs that adhere to public specifications such as FIPA.
MAS *applications are complex* to develop and it is *hard to debug* identify the source of errors: in the MAS vs. in the application.	Use an open-source MAS tool-kit, such as JADE, that has tools to help develop the application.
MAS communication involves *complex multi-layered application protocols*. These are complex to create and to parse.	Trade-off the expressivity of different message syntaxes and semantics vs. ease of use.
Throughput can be poor.	Reuse existing non-agent protocols as much as possible. Do not make everything an agent, selectively use agent communication.
MAS applications are complex to *maintain*.	Do not make everything an agent. Put agents on the fringes, not in the core of the system. Use as and when necessary.
It is difficult to agree a *semantic message model* and if one is agreed, it is hard to not to change it.	It may be impossible to agree and permanently freeze an ontological model for interaction. Develop a MAS model that can handle a living ontology. Develop ontology services that can detect and resolve ontology inconsistencies.
MAS middleware and services are immature and lack scalability.	Use stable conventional middleware and services to underpin or as a substitute for agent services.
MAS message transport services may not be optimised for non-agent communication.	Use multiple transport services: agent and non-agent and bridge between them.

5.9 CONCLUDING REMARKS

This chapter has given an overview of database integration techniques. The approach adopted for the EDEN-IW prototype is a hybrid approach, using where possible conventional Web and database approaches supplemented by the use of agents. This approach has the important advantages of being simpler and easier to maintain than a pure agent-based approach. For example, the agent interface to database resources is based upon standard Java Web and database APIs such as servlets and JDBC. The directory service uses an underlying Web-based service architecture.

Agent interaction is used to support a richer (semantic) interaction and information exchange, where necessary. Agent interaction forms the backbone of the whole system to allow all the different parts to interoperate, and to support user query transparency.

Chapter 6

SECURITY ISSUES FOR DISTRIBUTED SOFTWARE AGENTS

A. Papageorgiou

6.1 INTRODUCTION

Currently data on inland water quality and quantity is available primarily in databases, which are only accessible for internal purposes in the national environmental ministries.

However, the need for making this existing environmental inland water data available to users (through an interface independent of the physical or logical location of the underlying databases) raises significant security issues. These issues need to be addressed in order to optimise the balance between full security – which can be basically only obtained by having isolated systems – and the drive for openness in a distributed world.

This chapter examines the use of the technology of independent software agents. Each agent has a specific task in the complex process of processing each request, which generally involves:

- the questions raised by the user
- the formulation of specific (but database-independent) queries
- specific queries sent to the databases, which appropriate agents then evaluate as potentially having a valid answer (or part of the answer) for the actual question.
- the execution of these queries in the database and the formulation and transmission of the results that are produced.
- the presentation of the retrieved data to the user.

The examination of the development of the software agent platform will be based on the FIPA standard and the associated open-source implementations (for instance, FIPA-OS or JADE).

This chapter addresses several security issues regarding the installation and functioning of the resource agent. Such an agent will be associated with a data source, typically a database system, and it handles the interaction with this data source in order to retrieve and make available data requested by other agents.

Section 6.2 discusses general principles about the security measures. These measures can be enforced in the RDBMS level and provide guidelines for the database owners.

Most RDBMSs provide means for ensuring data security and integrity. The implementation of these means depends on the RDBMS vendor, though the general database security principles should apply for most RDBMS products. Section 6.3 discusses issues relevant to the interfaces between data sources and agents. Since the development of EDEN-IW will be based on a FIPA-compliant implantation, the discussion is concentrated on the Java Data Base Connectivity (JDBC) Application Programming Interface (API).

Since software agents and the databases will typically reside in private networks protected by firewalls, Section 6.4 briefly presents the various network configurations used in corporate environments.

Section 6.5 discusses the resource agent installation scenarios, the impact such an installation may have on the hosting network, and the software requirements in case the resource agent is installed outside the organisation network.

6.2 RDBMS SECURITY MECHANISMS

Independent of the installation of the resource agent, the security mechanisms offered by most RDBMSs should be used in order to enforce the proper security policy.

The following mechanisms will control access to the data held in a RDBMS.

6.2.1 A Privileges and Roles Scheme

A privilege can be either a *system privilege* or an *object privilege*. In the former case, the user is given the right to execute a particular type of SQL statement (e.g. create a table or connect to the database). In the latter case, the user is given the right to access another user's object (e.g. select rows from another user's table or execute another user's stored procedure).

Roles are named groups of related privileges that can be granted to users or other roles.

The use of roles provides:

- Reduced privilege administration, since rather than granting the same set of privileges explicitly to several users, this set of privileges for a group of related users can be granted to a role. Then only the role needs to be granted to each member of the group.
- Dynamic privilege management, since if the privileges of a group must change, only the privileges of the role need to be modified. The security domains of all users given the group's role automatically reflect the changes made to the role.
- Selective availability of privileges, since a role can selectively be enabled or disabled for a user. This allows specific control of a user's privileges in any given situation.

The above scheme can be utilised to create and manage roles specific to the function of the data that is made available.

6.2.2 Views

A view is a tailored presentation of the data contained in one or more tables (or other views). It takes the output of a query and treats it as a table; therefore, a view can be thought of as a "stored query" or a "virtual table".

Views provide a means to present a different representation of the data that resides within the database tables. Therefore, they can offer an additional level of table security by restricting access to a predetermined set of rows and/or columns of a table.

6.2.3 Authentication/Access Control Mechanisms

The authentication and access control mechanisms provided by an RDBMS can also be used. These authentication mechanisms, depending on the RDBMS vendor, can include the username-password schema and authentication using digital certificates, and can restrict access to the database in terms of combinations of user-IP address.

6.2.4 Auditing/Logging Mechanisms

Most RDBMSs provide facilities for logging and auditing user's activities, as well as for generating relevant reports.

6.3 INTERFACES BETWEEN AGENT SOFTWARE AND THE RDBMS

The communication between a database and a software component is implemented using middleware software that provides the database connectivity, such as JDBC and ODBC drivers. This document will concentrate on JDBC technology, since the FIPA standards use Java. JDBC provides cross-DBMS connectivity to a wide range of SQL databases, as well as access to other tabular data sources, such as spreadsheets or flat files. The JDBC classes are modelled after ODBC. They provide standard features such as: simultaneous connections to several databases, transaction management, simple queries, manipulation of pre-compiled statements with bind variables, calls to stored procedures, streaming access to long column data, and access to the database dictionary and descriptions of cursors. JDBC supports both a static and a dynamic SQL.

The JDBC (JDBC-SUN, 2005) technology drivers fit into one of the following four categories, as shown in Figure 6.1:

1. *JDBC-ODBC bridges*, which provide JDBC API access via one or more ODBC drivers. Note that some ODBC native code, and in many cases native database client code, must be loaded on each client machine that uses this type of driver.
2. *Native-API partly Java technology-enabled drivers*, which convert JDBC calls into calls on the client API for Oracle, Sybase, Informix, DB2, or other DBMS. Note that

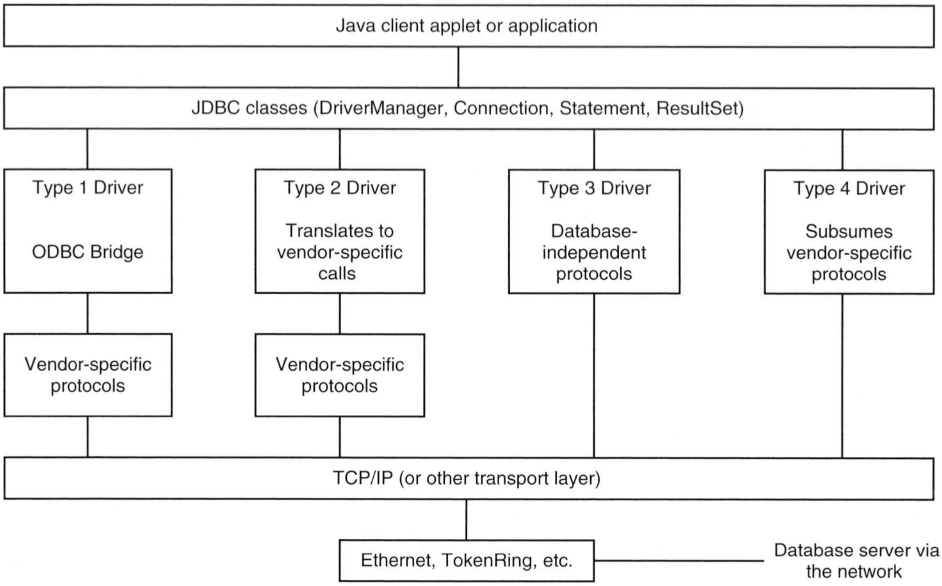

Fig. 6.1. JDBC drivers.

like the bridge drivers, this style of drivers requires that some binary code be loaded on each client machine.

3. *Net-protocol fully Java technology-enabled drivers*, which translate JDBC API calls into a DBMS-independent net protocol, which is then translated into DBMS protocol by a server. This net server middleware is able to connect all of its Java technology-based clients to many different databases. The specific protocol used depends on the vendor. In general, this is the most flexible JDBC API alternative. It is likely that all vendors of this solution will provide products suitable for Intranet use. In order for these products to also support Internet access, they must handle the additional requirements for security access through firewalls, etc. that the Web imposes. Several vendors are adding JDBC technology-based drivers to their existing database middleware products.
4. *Native-protocol fully Java technology-enabled drivers*, which convert JDBC calls into the network protocol used by the DBMS directly. This allows a direct call from the client machine to the DBMS server and is a practical solution for Intranet access. Since many of these protocols are proprietary, the database vendors themselves will be the primary source for this type of driver. Several database vendors have these in progress.

A discussion about the suitability of each of the above four categories of JDBC drivers in the EDEN-IW context is made in Section 6.5.

6.4 NETWORK CONFIGURATIONS

In an ideal world, all the software components in a distributed system would be able to communicate directly with each other and with the databases involved. However, in the real world, we must take into account certain security-related restrictions.

The various components of the application will be in firewall-protected networks. This section briefly presents the most common firewall mechanisms.

A firewall can be:

- A packet-filter
- An application level gateway
- A combination of the above.

The sections that follow describe these firewall configurations.

6.4.1 Packet Filter

Figure 6.2 depicts the packet filter firewall configuration. Packet filtering is the selective routing of packets between internal and external hosts. It can be done by most of today's routers, as well as dedicated firewall software or kernel modules (e.g. Linux's IP chains). The function of packet filtering is to either allow or block certain types of packets in a way that reflects the security policy of a network. The packet filtering routers are also known as "screening routers". An ordinary router, however, looks at the destination address of each packet and figures out which output interface to send the packet to, based on a routing table, while a screening router does the same sort of route determination, but also decides whether the packet should be routed or discarded.

The routers filter the incoming TCP/IP packets in terms of:

- IP source address, IP destination address
- TCP/UDP source port, TCP/UDP destination port

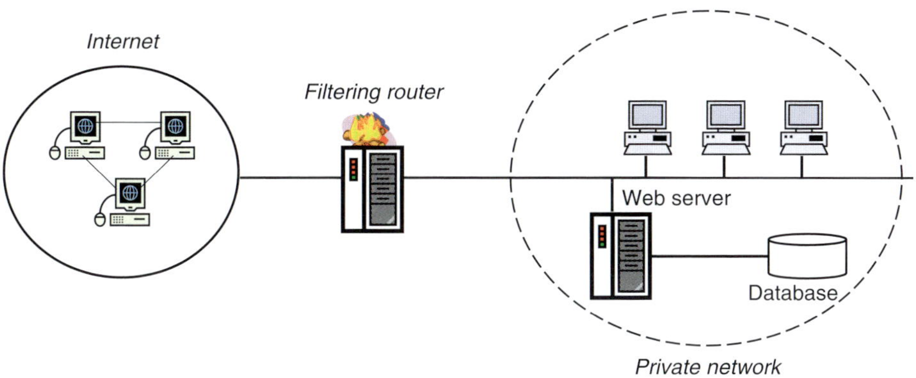

Fig. 6.2. Packet filter.

- Protocol (TCP, UDP, ICMP, etc.)
- ICMP message type.

6.4.2 Application Level Gateway

An application level gateway is a specialised application or server program that runs on a firewall host. This machine is known as a "bastion host" – a system that is specifically made secure for use in a firewall host, and its IP address is the only IP address from the organisation's network that is available from the Internet. These machines are generally dual-homed (i.e. they have two network interfaces, one for handling the communication with the public network and one for handling the communication with the private network), so that packets from the public network cannot flow directly to the private network. In the configuration shown in Figure 6.3, the router permits only traffic from/to the bastion host.

An application level gateway generally provides a replacement connection for the actual service (e.g. a reverse proxy server that serves as a Web server), and is capable of inspecting the data as well as the packets. Hence, it can keep track of the state of the communication and validate that the protocol conforms to the organisation rules. Furthermore, all outgoing requests (e.g. Web access) from the private network are first sent to the application gateway. The application gateway can log the address of the requesting machine and it can restrict access to certain sites or protocols (e.g. HTTP, FTP, telnet).

Application-level gateways are often known as proxies or proxy services.

6.4.3 Screened Subnet

The network configuration most commonly used in corporate environments is depicted in Figure 6.4.

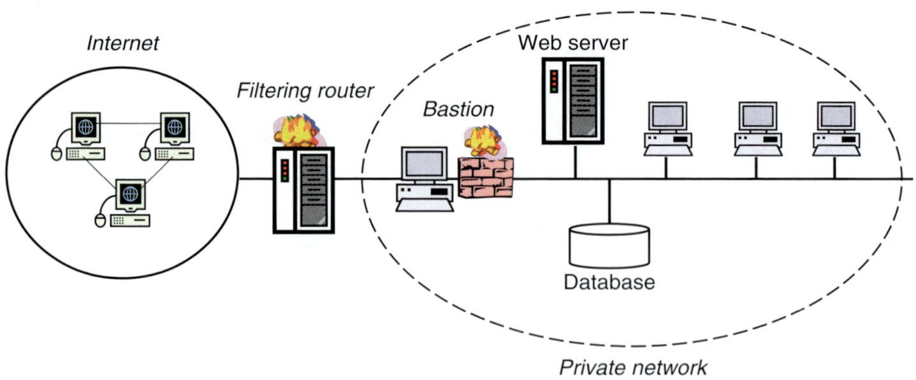

Fig. 6.3. Packet filter with bastion host.

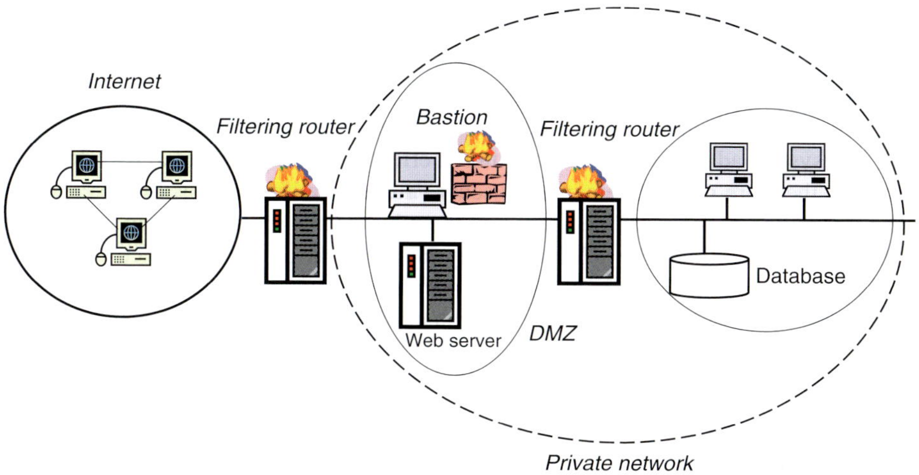

Fig. 6.4. Screened subnet (DMZ).

In order to achieve a greater degree of protection, externally accessible machines are placed on a separate network. Such a network is known as a "perimeter network", or "demilitarised zone" (DMZ).

This design consists of two screening routers: one between the public network (Internet) and the DMZ and one between the private network and the DMZ. An attacker would have to penetrate through both routers to get to the internal systems. Even if an attacker succeeds in penetrating a service running on a bastion host, he/she will not be able to see packets on the internal network.

The key filtering rules are:

- For the external router
  - disallow packets from the Internet that masquerade as packets from the internal network or the DMZ
  - disallow packets that are not destined for a DMZ machine
  - allow only packets destined for allowed services on the DMZ.
- For the internal router
  - allow only packets that originate from the DMZ network.

6.5 INSTALLING THE RESOURCE AGENT IN A CORPORATE NETWORK CONFIGURATION

Two possible scenarios can be identified for the installation of the resource agent.the software agent resides *inside* the private network and the software agent resides *outside* the private network.

traffic from the hosts of other agents, although this may require a frequent update of these rules, since the EDEN-IW application will evolve and new agents will be added.

If the network configuration uses a bastion host, then two options can be identified:

1. *Install the resource agent in the bastion host* Since the other EDEN-IW agents must locate the resource agent and none of the internal network's IP addresses should be visible from the public network, the resource agent must be installed on the bastion host.
2. *Use RMI proxy services* This option requires the development or purchase of software that will be installed on the bastion host and provides proxy services for the RMI servers that reside inside the protected network. A commercial product that provides proxy services for RMI clients and reverse proxy services for RMI servers is RMI proxy (RMI Proxy – Telekenesis, 2005).

Resource agent implemented as a Java servlet

In this scenario, it is assumed that the organisation uses a typical firewall configuration that blocks all network traffic to the private network, except traffic to ports used for HTTP and HTTPS. The software component, which will act as a resource agent, is developed as a Java servlet. Servlets (Java Servlet – SUN, 2005) are server-side mechanisms for receiving and responding (among other things) to HTTP requests. Like applets and EJB components, a servlet cannot operate on its own – it only operates in the context of a servlet container.

Without the servlet container, the servlet could not run. As specified in the Java Servlet 2.2 specifications, the servlet container works in conjunction with a Web or application server. They provide network services over which decode requests and format responses are sent.

Figure 6.6 illustrates how the servlet container fits into the company network and how the servlet fits into the servlet container.

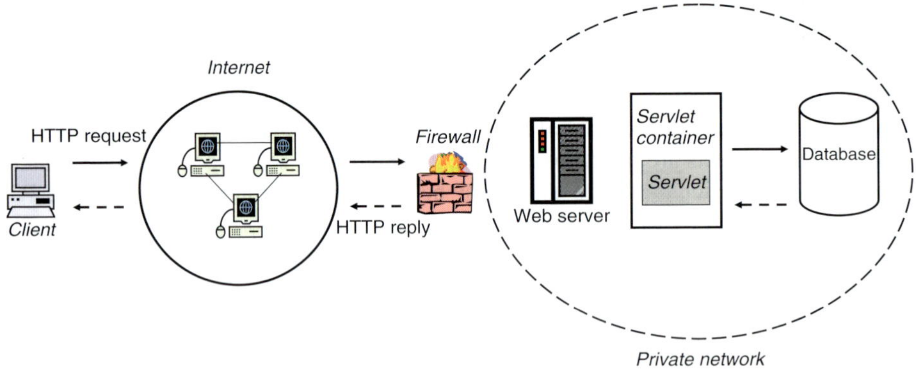

Fig. 6.6. Servlet approach.

Any client that wishes to interact with the database will have to issue an HTTP request to the company's Web server. The Web server then directs the request to the servlet container, which invokes the appropriate servlet. The servlet will interact with the database, and the Web server will reply with an HTTP response that contains the produced results. In order to ensure that the HTTP transaction is secure, Java Secure Socket Extension (JSSE) (JSSE-Sun, 2005), provided by SUN, can be used. The JSSE is applied on the client side in order to accomplish HTTPS over the Web.

RMI vs. Java servlet

Although this alternative approach is out of the FIPA context, it presents certain advantages over the RMI server approach:

- RMI communicates through Java serialised objects. Even though any data can be encapsulated into a Java object, HTTP supports RMI directly.
- The Java client can talk to other non-Java applications that already reside on the server, such as CGI scripts.
- The servlet container works within the typical business infrastructure, requiring little additional work from the system administrator (apart from the installation and configuration of the servlet container).
- Java security is different from Web server security. The security model and implementation of RMI are based on Java technology. Therefore, the administrator is faced with having to learn a new set of skills for RMI.
- Because HTTP tunnelling comes through the Web server into the servlet container, client requests will generally show up in the Web server log file. Open-source and commercial tools exist for reading Web server log files. These tools are useful for generating usage reports, and therefore are a handy way of monitoring application activity.

6.5.2 *Resource Agent Outside the Private Network*

When the resource agent resides outside the private network, and has to access a database inside the private network, it is necessary to apply a network-centric type 3 or type 4 JDBC driver (see Section 6.3).

- Type 4 JDBC drivers translate JDBC calls to DBMS-specific calls at the client level. Drivers of this type use no middleware server, since the client connects directly to the database server. The use of this type of JDBC driver assumes that the RDBMS vendor provides one, and that the firewall used to protect the network has knowledge of the protocol used. Furthermore, for the communication to be secure, either the driver must offer encryption capabilities or the database must support encryption natively.
- Type 3 JDBC drivers provide access to a database via a three-tier architecture consisting of the JDBC client (the recourse agent), the JDBC server, and the database being accessed. A type 3 driver and its associated middleware server adds an extra layer of security: instead of the resource agent logging into the database directly, as with type 4 JDBC drivers, the authentication process will be performed by the middleware JDBC server.

The middleware server can be placed on the bastion host, so that it can establish communication to both the protected network and the outside world. Thus, the resource agent can still indirectly reach the company's database server. The JDBC driver communicates with the middleware, and the middleware interacts with the database server that is inside the firewall-protected private network.

An important security advantage that some middleware solutions offer is encrypted connections between the JDBC client and the server. Because many of today's database vendors do not offer any way to transfer data over the network securely from the database server to a client, even for the login phase of the session, a middleware server, which implements encryption, may be the only secure option. Several JDBC middleware solutions use technologies such as SSL to encrypt the login phase of a connection as well as the whole session.

Additionally, middleware can unload the duty of connection management from the database server and it may offer more flexibility. Along with maintaining their own password lists, many middleware solutions can restrict the operations a JDBC client can perform on the database. Thus, a resource agent can be granted only *select* privileges for a particular view of the database from a specific network location in the middleware level.

6.6 CONCLUSIONS

The installation of a resource agent in a private network with the database providers is an issue that involves

- the RDBMS used by the participating organisation
- the organisation's network infrastructure
- the organisation's security policy

Although a separate case study must be conducted for each organisation, two general rules apply:

1. The servlet-based approach, or the installation of the resource agent outside the network, can serve the necessary purposes if the organisation is unwilling to either
   - change its network infrastructure (e.g. install the agent in its bastion host or acquire software that will provide RMI reverse proxy services) or
   - relax its security policy by opening a port that will be used by the resource agent
2. If the resource agent is installed outside the private network, then it must be made sure that either the available JDBC drivers or the RDBMS itself provides the security needed for the application.

PART II

THE EDEN-IW PROJECT

Chapter 7

INTRODUCTION AND OVERVIEW OF RESULTS

P. Haastrup

7.1 INTRODUCTION

The ambitious objective of the Environmental Data Exchange Network for Inland Water (EDEN-IW) project was to make existing distributed environmental data available to citizens and other users through an intelligent interface acting as a one-stop shop. The users, who may also be public authorities (for instance, environmental regulatory agencies) and the public, will be able to address their needs for Inland Water data through one common interface. This interface will be independent of physical or logical location of the databases, the database languages used, or the specific nomenclature used in a specific database. The usage will be possible, without knowing which database or databases contain the relevant information, by data mining the most complete set of information sources.

The project was conceived in early 1999. A significant effort went into developing the ideas and the subsequent project proposal, which was delivered to IST programs in June 1999, using new informatics tools allowing for online network submission of proposals (a big news at that time). The initial proposal included partners from the United States, to assure that the significant scientific efforts which had already been carried out through collaboration among US agencies (EPA, DOE, DOD, US-GS, and others) with the company MCC as main software developer could benefit the EDEN-IW project.

The base idea was to develop a system based on the technology of independent software agents. Each of the agents has a specific task in the complex process leading from the question raised by the user, through the formulation of specific (but database-independent) queries, through to the specific queries sent to databases which the agents evaluate as potentially having an answer or part of the answer for the actual question. Such requests for information require a common "language". Thus, a list of accepted and well-defined words is needed, i.e. the basis for an ontology relevant for inland waters.

When results are being returned, post-processing will be performed in order to supply the user with information in the most useful form. This post-processing consists of harmonising, aggregating, and presenting them in a consistent form, allowing variation in the level of detail presented, and integrating decision support tools for environmental management for the benefit of policy makers.

7.2 MAIN RESULTS

A number of results have been obtained during the project, and a prototype is now available.

The progress of the project has been reported thoroughly, both as external reports and as contributions at public conferences. One example is given in Haastrup and Würtz (2002).

The prototype, where independent software agents operate on a limited number of databases and in a limited number of languages, shows the value of various advanced information technologies for environmental data exchange and thereby eases environmental reporting and planning.

Data from more than one database can be interrogated at the same time, and the results elaborated in one operation in the EDEN-IW prototype, thus, demonstrated the value of the agent technology and the opportunities for linking other data resources into the system. This and other features are shown in the screenshots from the current prototype (Figures 7.1–7.3). A detailed description of the user interface is given in Chapter 12, and the connected map service is described in Chapter 13. The performance of the prototype is investigated in Chapter 14.

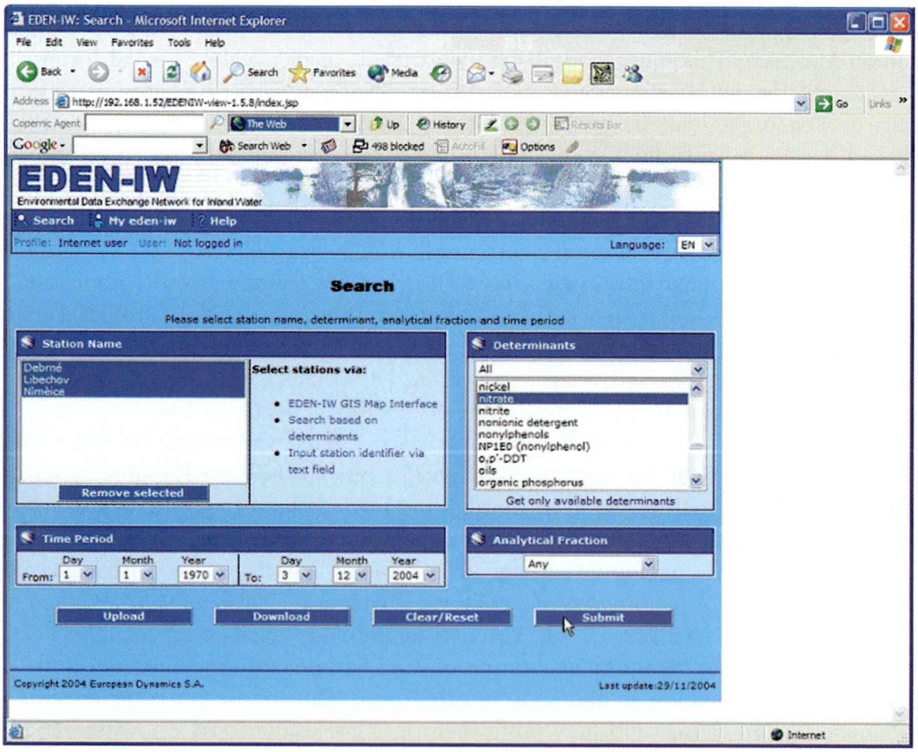

Fig. 7.1. EDEN-IW main search panel.

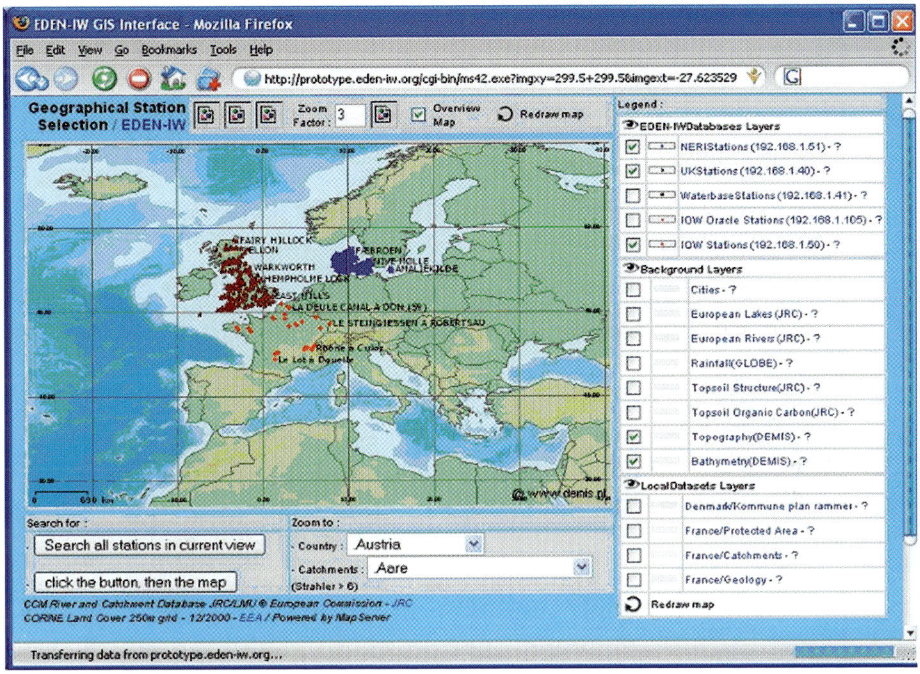

Fig. 7.2. Station search via map interface.

7.3 INDEPENDENT SOFTWARE AGENTS

As already mentioned, the EDEN-IW system is based on the technology of independent software agents (see Figure 7.4). This implies that each of the agents has a specific task in the process leading from a user query, through the complex process of providing the best possible answer. The processing of the requests requires a common "language" of well-defined keywords, which is collected in an ontology.

The implementation was made in an open-source environment, the FIPA (Foundation of Intelligent Physical Agents) standard (www.fipa.org), facilitating later applications in other areas. The JADE (Java Agent Development Platform) was chosen as the actual implementation of the FIPA specifications. JADE (http://jade.cselt.it/) has shown itself to be robust and to be supported via documentation, tutorials, tools, deployment, and examples.

In Figure 7.5, the user agent is the first to process the request from the user for specific information. Depending on the situation, the request may then need to be translated through the ontology service before it is passed to a directory (broker) agent and other agents, which assure that the request is forwarded to the relevant databases. At the other end, the resource agents, which are specific to each of the databases, are individual pieces of software. They are installed in the same computer as the database (resource)

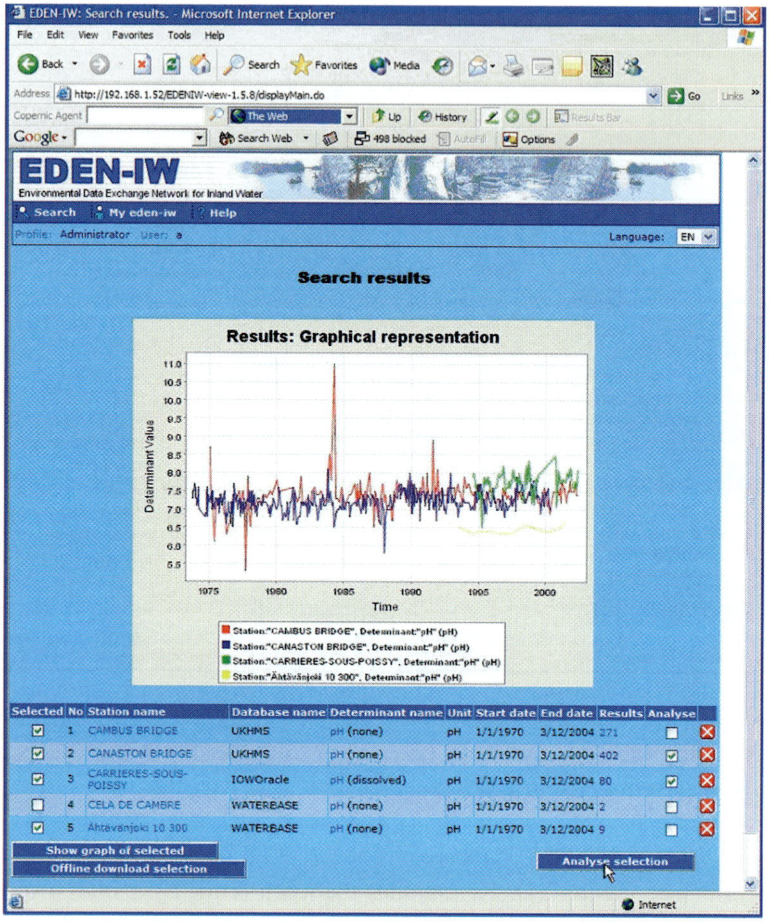

Fig. 7.3. Graphical representation of results from several databases.

in question, acting as data portals. On the way back to the user, the replies again may (or may not) pass the translator (ontology) service to furnish a multilingual output as required.

The software agents interact using a generic multi-level application protocol based on speech acts called Agent Communication Language (ACL). Although many services with the EDEN-IW system will be accessible via the agent interface, some lower level services are available via a non-agent interface such as a Web service.

There are four types of agents in the system: user agent (UA), directory agent (DA), task agent (TA), and resource agent (RA) (see Figure 7.6). Non-agent components include ontology services, Web services, and data resources. A detailed description of the agent system is given in Chapter 8.

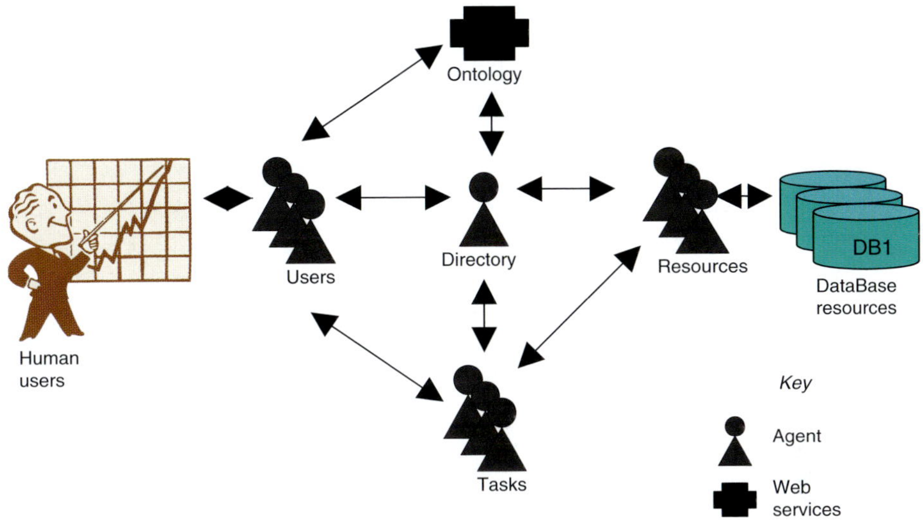

Fig. 7.4. Conceptual model of the various software agents in the EDEN-IW system.

7.4 INLAND WATER APPLICATION

The research and development of a specific application for Inland Water data based on the software agent technologies mentioned above has produced the EDEN-IW prototype. The prototype connects a number of relevant databases to the system.

Data from more than just one database can be interrogated at the same time, and the results elaborated in one operation, in the EDEN-IW prototype, thus demonstrating the value of the agent technology and the opportunities for linking other data resources into the system.

7.4.1 Databases Connected

The resources to the prototype comprise data from four different existing databases (see Figure 7.7). The databases contain a number of common parameters that can be applied by EDEN-IW. Data from the databases has as far as possible been maintained in the original table structure. The databases have been documented in local view ontology files and mapped to the global ontology. Each of the databases has contributed to the understanding of the challenges in the integration of heterogeneous databases. The project has concentrated on river-monitoring data. The databases originate from three national and one international sources:

The three national databases contain individual observations of the determinants, whereas Waterbase contains values, aggregated over time periods (Annual, Summer, and Winter). The values comprise various statistical values: mean, minimum, maximum,

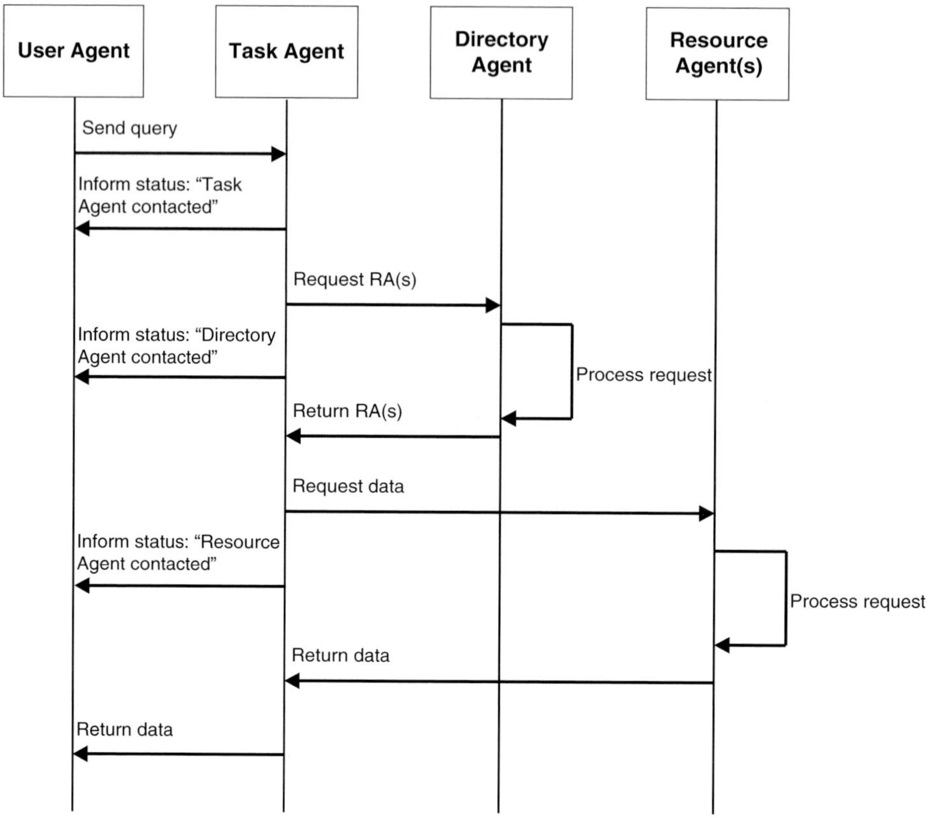

Fig. 7.5. Handling of queries.

The User Agent supports a Web-based User interface for the user and is the first and last point of entry to access the facilities of the internal agent-based EDEN-IW system (see Figure 7.4) Initially it receives the query from the user interface; it then transforms this into RDF format and passes it the Task Agent. The interaction between the Task Agent and the Resource Agent will provide the former with the data satisfying the query to pass back to the User Agent.

The Task Agent is the coordinate of the multi-agent system, and the manager of the agent interaction. When it receives a query from the user agent, it first analyses the query, then different plans are made according to different queries.

The Directory Agent is the central repository for metadata. It stores a list of all available stations and active resource agents. It also holds a list of all determinants available at a particular database.

The Resource Agent provides the resource wrapping service to access conventional databases to retrieve the IW data. The wrapping service accesses both EGV (EDEN Global View) and LDV (Local Database View, see details below) ontologies via appropriate ontology "adaptation" services.

Fig. 7.6. EDEN-IW agents.

IOW: International Office for Water, France – The French national monitoring database.
NERI: National Environmental Research Institute, Denmark – The Danish national monitoring database.
UK-HMS: Environment Agency for England and Wales, England – The national Harmonized Monitoring Scheme database.
WATERBASE: European Environment Agency, Topic Center for Water – The European database of selected aggregated monitoring data.

Fig. 7.7. The databases connected to the EDEN-IW prototype.

median, standard deviation, 10% percentile, and 90% percentile. The Waterbase data originate from and cover 27 countries with a total of 3438 river stations.

The databases are mostly available as Microsoft Access files. Tests have been performed with the databases in three other database engines, Microsoft SQLserver, Oracle 9i, and Oracle RDB, and in two cases minor modifications to the original structure have been necessary.

The pilot data set covers a total of 69 determinants. The number of determinants mapped in each database varies between 57 and 19. Eight determinants are present in all the four databases.

7.4.2 Ontologies

For the use in the prototype, an *Ontology* covering the inland water domain has been developed, permitting queries to the four databases, which are integrated and mapped against each other. The Ontology covers a significant number of water-related indicators and concepts, with a multilingual user interface, though queries in the current prototype are limited to the use of a single language at a time.

During the project, it became clear that various databases use similar concepts for non-identical information. Therefore it was not sufficient to examine the issues linked to conversion of, for instance, units (where one database may report a given concentration in the unit mg/l, while another may use another measurement unit), and database codes. The underlying definition of the concepts reported in the database also had to be examined in detail, in order for the data points to be brought into a comparable form. Ontologies had to be examined in considerable details, in order to avoid aggregating results that were not comparable and where a comparison could lead to misinterpretation. The conceptual notions in the water quality domain (such as parameters and analytical fractions) are not universal.

The EDEN-IW conceptual ontology approach is based on a separation of global knowledge (EGV – EDEN Global View) from the more database-specific local knowledge (LDV – Local Database View ontology). The approach opens up the opportunity to exploit the union of knowledge and data. The EDEN-IW system is able to do this by using a semantic model that is open and scalable such that new determinants and concepts can be added, new applications can be integrated, and new specialised user interface can be connected without undue effort.

The EDEN-IW ontology model and associated applications are represented using the W3C RDF-, DAML- and OWL-based models (see DAML example in Figure 7.8) and implemented using the FIPA agent forum standards allowing agents to parse messages. Queries are expressed in the EGV ontology. These are then mapped into the LDV models of the corresponding database resources and then to a database query. The same agent services also translate the responses from the database back into to the common language (EGV) terms. A detailed description of the ontologies applied for connecting the databases is given in Chapter 9.

7.5 ENVIRONMENTAL ANALYSIS

The initial scope for the research into environmental analysis was very broad. The initial main goal was to enrich the possibilities for an environmental manager to process and use the data in a meaningful way. It was subsequently narrowed to an implementation along the lines of an end user, which could be defined as the "Technical Manager of a river basin".

Here a key question of interest could be of the type "What (which river branch) should I clean to get the best results for the local population?"

Starting from this technical level, an implementation of agents taking care of operations like scaling, aggregating data, presenting and projecting time, and presenting uncertainties was made. This is shown in Figure 7.9.

It was found that bringing the relevant functionality closer to the user, by implementing the functionality in the User Agent, allowed a series of benefits, among them the possibility to elaborate the results of earlier user queries in the system, together with new and more specific searches and results from the data bases.

As already indicated, uncertainties fall in a number of categories, some of which are linked to the data, to the sampling methods, and to the analysis, while others are subsequent methodological issues (types of mean values, for instance). It is important to recall that the issues of uncertainty and thus of the quality of the data are strongly linked to the *use of the data themselves.* Thus, data of a given uncertainty may be good quality data for one application whilst at the same time insufficient for another application. The uncertainty in the data thus has to be represented in such a way that the user can understand if the actual quality is (or is not) sufficient for the application at hand. On this point, the current prototype is not very strong, because the basic quality information linked to the data points is difficult to distinguish from the variation in the data.

However, with the analysis tools now integrated in the prototype (described in Chapter 12), additional functionality can be added relatively easily for potential users afterwards.

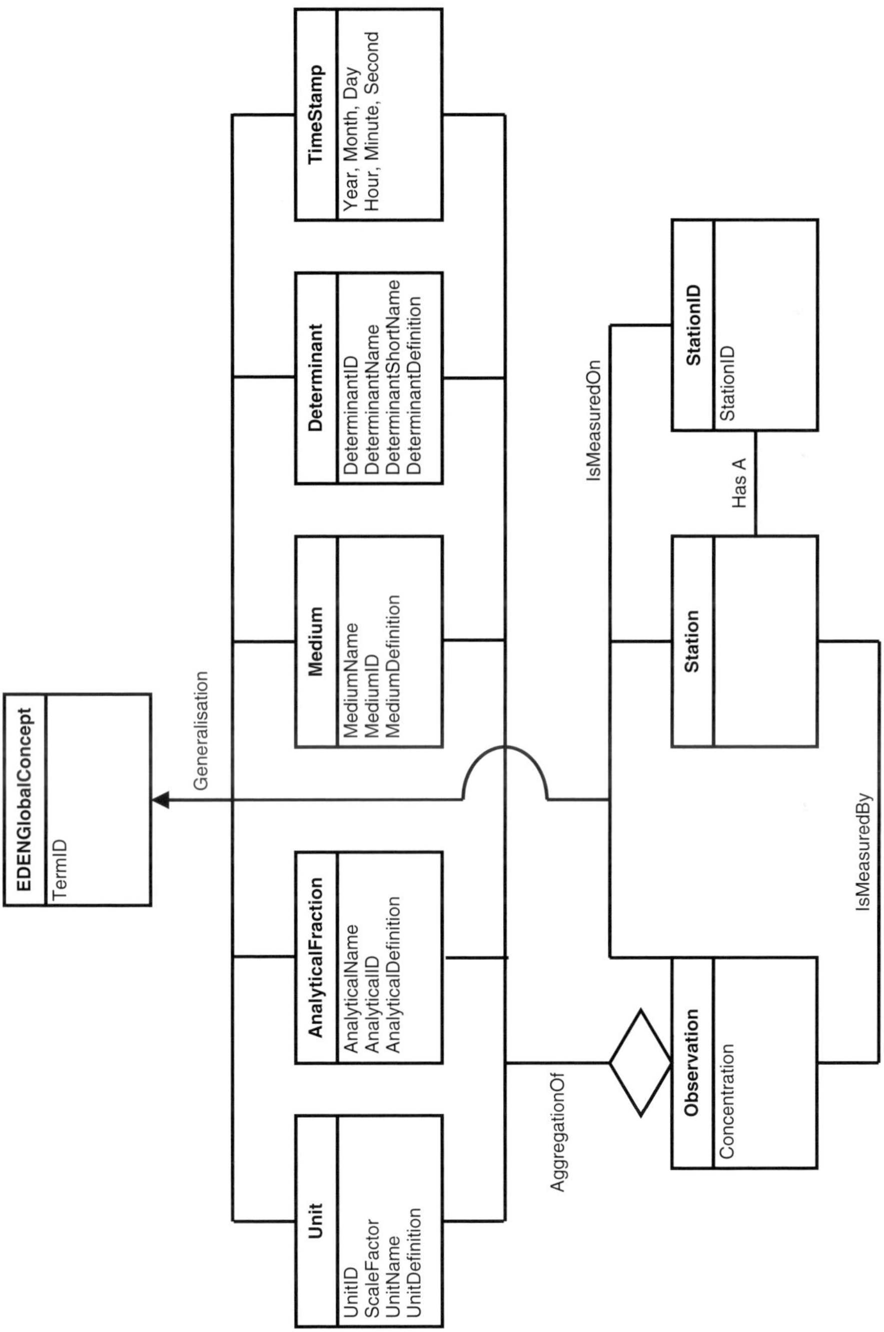

Fig. 7.8. UML diagram of DAML global view.

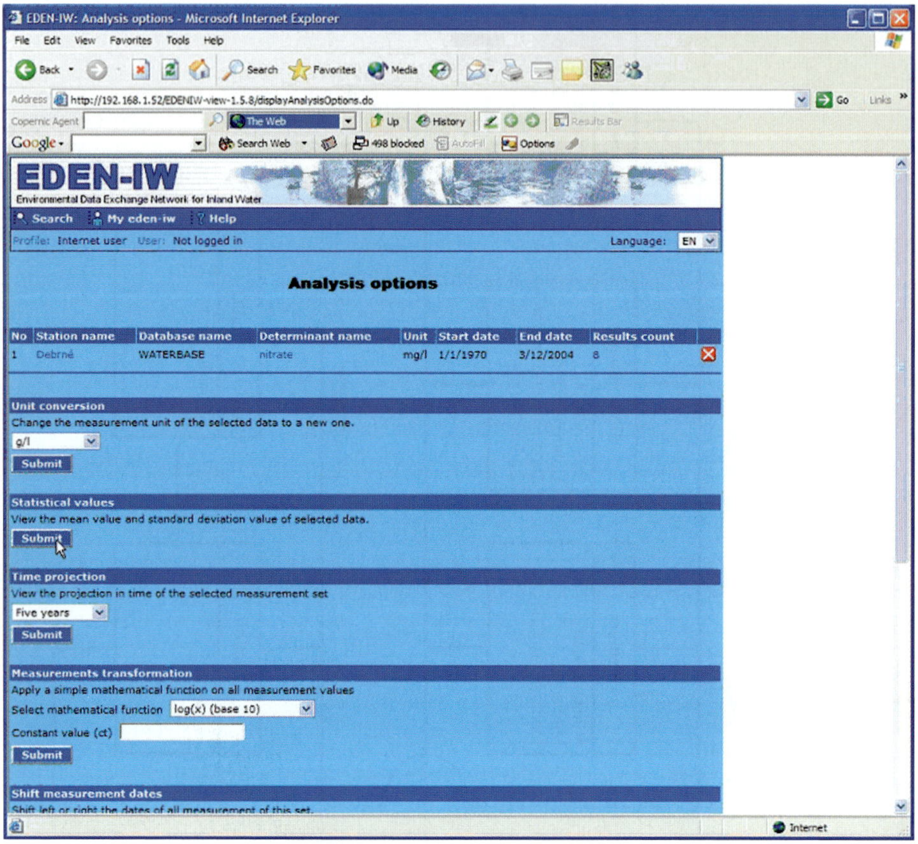

Fig. 7.9. Analysis options panel.

7.6 AGENT COMMUNICATION VISUALISER

In order to look behind the scenes of the events in the agent system, an agent communication visualiser has been developed, capable of showing the flow of messages between the various agents in the system, thus giving a realistic and online view of events. The visualiser provides an easy and configurable Web interface for viewing messages between agents. The software includes American English, French, and Danish menu labels.

The positions of the agents and services are identified on a catchment-basin map of Europe, along with explanatory legends that identify the types of EDEN-IW agents and the messages that pass between them. Messages are displayed as different-coloured shapes that travel along paths between agents.

The visualiser is designed to support multiple ways of viewing agent communication. It can be configured with different icons for agent and message types. The display

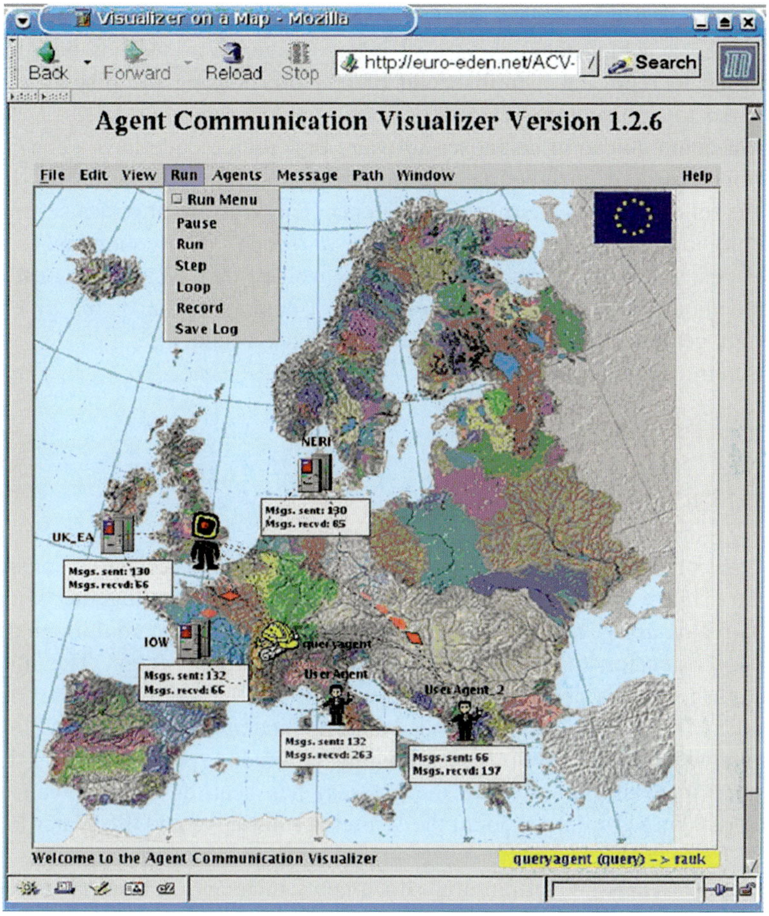

Fig. 7.10. Agent Communication Visualizer with map of Europe.

background can be altered as well, so that the message traffic pattern between a certain group of agents can be viewed in more than one way.

In Figure 7.10 a message stream is displayed on a background of a map of Europe. A red diamond indicates the return of a reply from the directory agent to the user agent located in Greece. Meanwhile, the task agent has just issued a request, indicated by the star, to the UK resource agent. The International Office of Water (IOW) resource agent is replying to the task agent at the same time. The Run menu is selected, showing some of the possible ways of controlling the display of the messages, including pausing the display, stepping one by one through the message log, or saving the log for later examination or replay.

Because computer network message traffic often occurs within time-spans shorter than can be comprehended visually by humans, the ability to display messages in a meaningful approximation to real time is important. For this reason, the applet also permits one to adjust the speed at which messages travel between agents, as well as how much time should be used to represent a millisecond of real time during the display of message traffic, in order to represent pauses between agent messages appropriately.

The agent communication visualiser software and source code have been made publicly available under an open-source license during the final months of the EDEN-IW project at the following URL: http://euro-eden.net. The Website provides several demonstration configurations of the applet, including a "space" demo showing some of the more sophisticated configuration possibilities, as well as the "cloud" demonstration and EDEN-IW demonstration using the Europe, and a demonstration of the pipeline order model using a genome sequencing pipeline for its background.

The Visualiser is described in detail in Chapter 10.

7.7 SEMANTIC MANAGEMENT

The main goal of this work was the assurance of efficient and effective semantic management. To achieve this goal, a specific Inland Water ontology, together with specific Inland Water glossaries, was developed. Two external reference multilingual thesauri, allowing easy extension to multi-language situations, support these. The initial effort includes English, French, and Danish.

During the EDEN-IW project, access is being facilitated to environmental information using a common controlled reference terminology, tailored to specific needs of the project. Both the controlled terminology and the glossary has been developed even above the level required, taking also into account the Waterbase initiative and the Water Framework Directive.

The series of EDEN-IW terminology documents includes the components shown in (Table 7.1).

At the onset of the EDEN-IW project, a unique, standard, multilingual reference system of terms of the Inland Water domain did not exist. In principles, the "languages" of the

Table 7.1. The EDEN-IW terminology components

EDEN-IW Glossary: a general purpose, reference background terminology ("language") of the Inland Water domain, to be used during the development of the EDEN-IW ontology system.
EDEN-IW Set of Terms: of a cross-terminological comparison between the IW databases of NERI and IOW.
EDEN-IW Terminology Reference System (TRS): the result of merging the Glossary and the Set of Terms terminologies, with a further addition of terms from the multilingual thesaurus EARTh and other pertinent sources.

IW domain encompassed the academic text of hydro-sciences, the internal languages of IW databases and ad hoc compiled glossaries, in different languages.

The overall Terminology Reference System for Inland Water (EDEN-IW TRS) consists of:

- The terminology of the Cross-Mapping Table, which is the core terminology of the reference system;
- The terminology of the EDEN-IW Glossary
- Pertinent terms from the EARTh thesaurus
- Synonyms and formulas of pesticides and of chemicals in general. These are obtained from several sources, and sources of sources of chemical information
- The terms of the list of Priority Substances of the Water Framework Directive
- Relevant terms from the text of the Water Framework Directive

Since a total equivalence in the three languages of the EDEN-TRS was not foreseen in the project, the multilingual list represents a subset of 1770 terms out of the total terminology in English, 2949 terms (see Table 7.2). This subset is restricted to the English terms having either a Danish or a French equivalent. The present state of the resulting multilingual lists is shown in Table 7.3.

The TRS is described in detail in Chapter 10.

7.8 DISCUSSION

The technologies explored and the results obtained, if more fully implemented than in the working prototype, enable the environmental manager to cross, compare, and use more than one environmental data base at any given time (the current situation) for the benefit of Europe's environment.

However, the results obtained in the EDEN-IW project are less spectacular than the initial hopes, and will require additional effort of consolidation.

Table 7.2. Equivalence between the languages in the TRS

Language	No. of Entry Terms/ Concepts	Total No. of Synonyms in the Complete Alphabetical List	[No. of Selected Synonyms in the Glossary]	No. of English US terms/ Variants	Total No. of Terms/ Concepts, including Synonyms
English UK	1866	1083	[159]		2949
English US	1866	1083	[159]	58	58
Danish equivalents	810	23	[23]		833
French equivalents	1350	94	[94]		1444
	Total: 5892	Total: 2283	Total: [435]		5284

Table 7.3. Number of terms in the multilingual lists

Number of Records	1770
English Entry Terms	1556
English Synonyms	214
English Total Terms	1770
Danish Entry Terms	798
Danish Synonyms	151
Danish Total Terms	949
Danish Missing Equivalents	758
French Entry Terms	1278
French Synonyms	210
French Total Terms	1488
French Missing Equivalents	278

Related to the *agent technology*, the agents and infrastructure produced can demonstrate to robustly support and enable integration of inland water domain knowledge, they are integrated in open-source architecture and they enable and support water domain knowledge from distributed heterogeneous sources. This is a significant and successful result of the research.

In addition an Ontology covering the inland water domain has been developed, permitting queries to four databases, which are integrated and mapped against each other. The Ontology covers a significant number of water related indicators, with a multilingual user interface, though queries in the current prototype are limited to the use of a single language.

Related to *Environmental Management*, the project has successful implementation of agents linked to operations like scaling, aggregating data, presenting and projecting time, and presenting uncertainties. The implementation is successfully done; scaling, aggregation, presenting and projection in time is possible; while uncertainty representation is still weak.

During the EDEN-IW project, access is being facilitated to environmental information using a common controlled reference terminology, tailored to specific needs of the project. Both the controlled terminology and the glossary has been developed even above the level required, taking also into account the Waterbase initiative and the Water Framework Directive.

In addition, the results of the EDEN-IW project have influenced the system design of the Water Information System for Europe (WISE) and the data system collecting information for the Intercalibration Exercise under the Water Framework Directive.

Support to the project by the database owners was also a difficult issue. In general, they welcomed the idea and accepted to describe the structure of their system. Nevertheless, they were more reluctant to be included in the prototype, for many reasons: security issues, risk of download of large volumes of data, risk of misuse of the results, and the workload to be dedicated to the project.

Chapter 8

MULTI-AGENT SYSTEM TECHNOLOGY IN DISTRIBUTED DATABASE SYSTEMS

S. Poslad, L. Zuo and X. Huang

8.1 INTRODUCTION

8.1.1 EDEN-IW System Overview

The EDEN-IW system provides high-level access to support the integration of heterogeneous distributed Inland Water (IW) databases. This is obtained by means of a shared semantic metadata model. The IW databases are heterogeneous with respect to their database schema, the multi-lingual terms, the DBMS vendor, the degree and type of metadata available online, and the semantics of the terms used.

Externally, the core EDEN-IW system appears to consist of two Web portals. The first is a user portal, where the user poses queries. The second is the data Resource Web portal that wraps the individual IW database resources (Figure 8.1). Internally the EDEN-IW system consists of a Semantic Web–type ontology model plus a Multi-agent System (MAS). The latter supports semantic data services and communication protocols.

At the user portal of the EDEN-IW system, example data and metadata queries that can be posed include the following:

1. Which are the Observation Values of Determinand X in Medium Y that have been measured at Station Z between time period T1 and T2?
2. Which are the Station Names where Determinand X has been observed with (or without) the value equal to (or less than, or greater than) a specified threshold?
3. Which are the Determinand Names available in the database X (Metadata use case)?

The reply to a query that identifies a given station can in turn be used to perform another query related to an observation value. The EDEN-IW system is able to handle information retrieval and integration from four heterogeneous databases: NERI, IOW, UK and EEA WATERBASE databases. The query answers are cached in the user interface. The system also allows the post-processing of the query results for comparison and analysis functions.

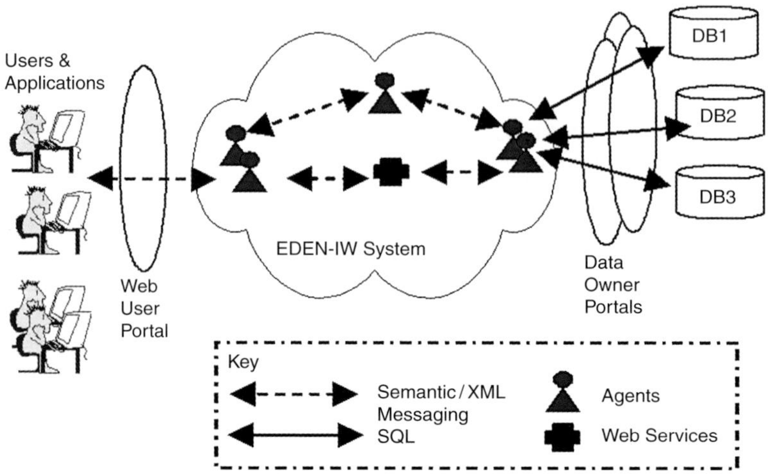

Fig. 8.1. The EDEN-IW system and its two Web portals.

8.1.2 Motivation for Using MAS

Previous work by other researchers using the InfoSleuth-based agent architecture (Nodine et al., 1999) in an earlier, related project has demonstrated the potential of multi-agent systems and semantic approaches to enhance environmental information retrieval. In contrast to the InfoSleuth approach, the EDEN-IW system adopted a more open system approach in terms of its use of specifications for the multi-agent systems and ontologies and in the way the semantic metadata architecture was modelled (Zuo and Poslad, 2003).

The main EDEN-IW system requirements are to support high-level queries in terms of query transparency and data harmonisation. To do this, it seeks to leverage two underlying technologies: ontologies and agents. First, a representation of database metadata in terms of an ontology model is used to classify domain concepts, properties, relations, and constraints. The use of an ontology model supports the exchange of machine-understandable structured data and automated processing, and enhances information queries and information searches. There are several potential benefits in using a semantic approach to enable information integration. Interoperability is eased, as there are likely to be fewer semantic differences than syntax differences. The underlying domain knowledge model is made explicit and is separated from the applications that create and manipulate it. Semantic processing by MAS supports the opportunity to expand a user query depending on the context (query augmentation); to integrate and aggregate the contents (content harmonisation); and to use the semantic model to classify, (re)structure, and to index information (see Chapter 5).

In order for the systems to be truly interoperable, more than a defined semantic data model is needed. Semantic services are needed to import, parse, and process semantic metadata instances, to map them to data resource instances, and to distribute and coordinate the metadata. Therefore, the EDEN-IW system is also based upon an MAS model. However, the use of the MAS model gives added benefits, beyond providing a semantic metadata processing and distribution framework. MAS communication is usually based upon an underlying communication protocol Speech Act theory that treats communication as actions (Searle, 1969). This gives a powerful approach to integrate human intentions and computation system service actions, and to communicate about processing and meta-processing (e.g. an analysis of the reasons to communicate in a particular way or to change the communication). Using this underlying model, MAS agents can coordinate messages, and process and reason about the semantic message exchange.

Later developments in speech act theory have included modelling the intentions of the sender in initiating communication. These can range from weak intention such as an intent to send a cry for help without any consideration of the receiver, to stronger intentions to initiate a specified reaction from the receiver to the still stronger intention of altering the beliefs of the receiver or for the receiver to take on board the beliefs of a third party (Ferber, 1999).

There are several other potential benefits of using MAS. They provide a rich set of messaging protocols to share and converse about the semantic model. The message protocols used are independent of the application domain and hence the same communication protocol or set of actions (such as send, acknowledge, refuse, and reply) can be used across applications. This supports a greater consistency and ease of invocation of common actions across multiple applications (Raskin, 2000). Hence, agents support a transparency and virtualisation notion because agents present a common set of communication actions to allow users to invoke different database resources and different data processes transparently. Agents can support the concept of dynamic virtual organisations, acting to a degree autonomously but organising themselves driven by the interaction context, e.g. requesters and suppliers could be organised and interact according to a master–slave relationship, or according to a market-place. That is, agents leverage the duality that exists between organisation and interactions, where an organisation is defined by the interactions it supports, and interactions do exist and are constrained with respect to a particular organisation (Ferber, 1999).

Multiple agents can flexibly solve complex information retrieval problems such as data harmonisation and aggregation from multiple data sources using autonomous, specialised agents that can coordinate their individual actions or complement each other to solve a problem: Agents support cooperative planning to coordinate the actions of others to solve a problem that they cannot solve alone. Agents can support multiple redundant plans, switching to alternative plans if one fails. Thus, they are offering support for fault-tolerance.

Agents can also act as powerful service mediators supporting flexible service requests to capability matches and isolating requesters from providers. They may, for example, act as a one-stop shop, thereby hiding the complexities of composite service invocation and interaction for the requesters, and thereby also providing privacy and impartiality to requesters and providers.

Table 8.1. Information retrieval application requirements and the corresponding agent properties that can be used to support them

Information retrieval application requirements	Corresponding agent properties
Usage transparency	Agent Communication is based upon an underlying communication protocol (speech act theory) of treating communication as actions that can integrate human intentions and computation system service actions.
Resource and service virtualisation–a set of universal service actions is supported	Agent Communication uses a common set of communication protocol actions used across all service instances.
Virtual organisation that is formed on demand to solve problems	Agents use plans to achieve goals, to coordinate actions of themselves and those of other agents.
Fault-tolerance	Plans can contain redundant plans; switching to alternative plans when one plan fails.
Protocols for semantic metadata/knowledge exchange	Common set of communication protocol actions supports knowledge exchange. The communication protocol provides a process context to interpret the content.
Dialogues such as flexible service request-provision mediation	Agent communication supports a rich set of dialogues such as contract-net, subscription, auctions, etc.
Reasoning about logically expressive (semantic) messages	Agents support reasoning with proposition, rules, and desired states.

Agents can reason about messages that contain logical expressions in order to provide the processing to support content harmonisation and to provide the flexibility to optimise the interaction according to the application context (see Section 8.1.2).

It is possible to highlight a set of specific useful properties of agents that are problem-driven (Table 8.1), i.e., support specific problem-driven requirements and provide a useful model to analyse a problem. This is in contrast to a tendency to introduce solution-driven models of agents, i.e., agents are potential generic technological solutions merely because they have a set of useful characteristics such as autonomy, proactivity, mental deliberation, and an ability to support rich coordination.

8.2 SPECIFICATION OF MAS USING FIPA STANDARDS

The motivation for specifying agent systems using the FIPA standards has been given in Chapter 5. To summarise, it is to promote MAS interoperability and to enhance maintenance of the design and implementation. Here the key specifications are described, and the way they are used is outlined. The Foundation for Intelligent Physical Agents, or FIPA, was formed in 1996 to promote the uptake of software agents in businesses at large (FIPA, 2005a, b). FIPA became an IEEE standards forum in 2005. The official mission statement of FIPA is:

> The promotion of technologies and interoperability specifications that facilitate the end-to-end internetworking of intelligent agent systems in modern commercial and industrial settings.

FIPA originally intended that specifications and standards encompassing both hardware (physical) agents, such as robots, and software agents would be developed, hence the use of the term 'physical' in the FIPA name. However, as FIPA progressed, the interest of the forum focused more and more on software rather than on hardware agents. A second key focus was on specification of the communication and interoperability between agents. This was preferred to standardising the specifications about how agents internally processed the received information. Thus, FIPA concentrated on standardising 'external intelligence' (rich interoperability) rather than on 'internal intelligence' (reasoning).

The first set of seven specifications was produced in 1997. This set included the following specifications: an Agent Management specification that originally defined the concept of an agent platform; and an agent communication language, and some applications such as travel assistance, network management, audio-visual entertainment, and personal assistance.

In the subsequent years, further specifications were added. The architecture and Agent Communication language have been refined, and the specification process has been formalised. Non-FIPA standards such as Web technologies have been increasingly integrated with the FIPA standards, and there is now more focus on abstractions and instantiations of Agent Communication for use in a heterogeneous world. The FIPA specifications are now the most commonly used design for MAS applications and have been deployed in a diverse set of domains.

8.2.1 Interoperability and Bootstrapping

It is highly desirable that agent specifications are sufficiently extensible and open to work in a heterogeneous and changing world. The extensibility and openness are desirable at two distinct levels of granularity. These are the agent level and the agent component level. The latter refers to the service components that underpin the agent.

At the agent level, if agent systems are to scale up in the market-place, leading to mass-market penetration, then openness allowing multiple vendors to freely add new agents and aggregate agents within a market-place, collaboratively, competitively, and dynamically is highly attractive (Poslad, 2000).

At the agent component level, it is desirable that the interface between the agent component and the agent does not bind the agent to a single particular instance of the agent component. For example, consider the agent transport: in early versions of the Agent Transport Specification, FIPA specified the use of a single, so-called base-line message transport – the Object Management Group IIOP transport. This was ideal for low volume transactions, wire-line and private networks without firewalls. However, if FIPA agents need to operate through firewalls, or to be used for high-transaction processing or for wireless environments, the IIOP transport protocol is not adequate. Agent component interfaces need to be more neutral and abstract – but yet be able to be grounded in specific technologies. For example, the current agent transport specification (and the Abstract Architecture specification) can be grounded in multiple message transport protocols such as IIOP and HTTP. The Abstract Architecture specification says that gateways are needed to interlink different types of non-agent component – but it does not deal in detail with bootstrapping. The specification of a global transport protocol – that must always be present – seems useful here in theory. This could then be used to negotiate the use of a more optimal transport protocol during a session.

8.2.2 Overview of the FIPA Specifications

A good overview of FIPA specifications is given in Poslad and Charlton, (2001). The FIPA specifications can be partitioned into five categories: Applications, Abstract Architecture, Agent Communication, Agent Management, and Agent Message Transport.

These specifications can then be used to help define the organisation or structure of an agent application as follows. The FIPA Abstract Architecture specification provides core abstract elements for modelling agent applications and services. The Agent Management specification provides some more concrete service elements derived from the Abstract Architecture elements for modelling standard agent services such as directory service and name service agents. The Agent Message Transport specification, Agent Communication specifications, and the Agent Management Specification define properties and (message-passing) interfaces for further standard agent services.

8.2.3 FIPA Agent Communication Specifications

The FIPA Agent Communication Language (ACL) specifications are based on speech act theory (Searle, 1969) described earlier. In practice, a FIPA ACL message, consists of (Figure 8.2):

- *A message header*: defines the transport details such as sender, receiver, and transport encoding.
- *Speech act or communicative act* (*CA*): defines a message type such as request, agree, inform, refuse, etc.
- *Dialogue or interaction protocol*: defines an interaction and process context for the individual messages (sometimes, the FIPA specifications refer to these as just 'protocols'), e.g. contract-net, Dutch auction, subscribed, etc.
- *Content language*: defines the logic and algebraic expressions (see below) to process the content of the message.
- *Domain ontologies*: define the semantics of the terms used in the content.

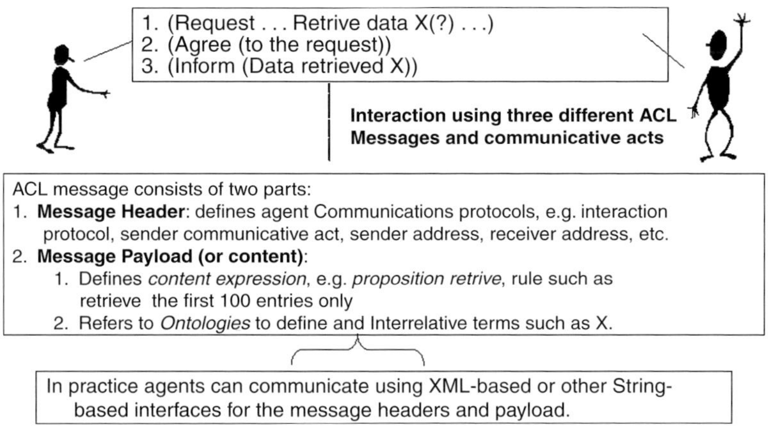

Fig. 8.2. Overview of agent communication in which one agent requests information from another.

- *A message body*: the content of the message expressed using the content language expressions and ontologies.

The use of the term 'Agent Communication Language' is multi-faceted: sometimes it is used to refer just to the communicative act part and sometimes to the combination of the communicative (also called speech acts), interaction protocols, and content for message transfer. The set of the order of 20 communicative acts forms the basic set of message types exchanged between agents. These communicative acts support several different purposes such as:

- *Information passing*: confirm, disconfirm, inform-if, inform-ref
- *Requesting information*: query-if, query-ref, subscribe
- *Negotiation*: accept proposal, cfp (call for proposal), propose, reject-proposal
- *Action performing*: agree, cancel, refuse, request, request-when, request-whenever
- *Error handling*: failure, not understood.

Often messages are more usefully modelled within the context of a dialogue or interaction: an exchange of several messages between the sender and receiver. For example, client-server systems can exchange messages within a request dialogue corresponding to sequences of communicative acts of request (e.g. then not-understood, agree or refuse, then inform or failure). The FIPA interaction protocols include: Request, Request-When, Query, Contract-Net, Iterated Contract-Net, English Auction, Dutch Auction, Broker, Recruit, Subscribe, and Propose. Domain-specific interaction sequences and extensions of communicative acts and interactions may also be defined.

Semantic content expression vs. semantic concept organisation

The content is the part of a message that represents the application domain–dependent component of the communication. It is worth noting that there is a difference in the way semantics are modelled and deployed in the W3C Semantic Web framework compared to the way it is modelled in the ACL-based frameworks. In the Semantic Web, there is a single representation of the application concepts, often organised as a tree or graph (weak Ontology) plus optional logic expressions that are found in the more expressive, strong ontologies. In the ACL framework, the models and representations of the application-specific conceptual model (or meta data) are modelled and represented separately (defined by the Ontology Language) from the application neutral, logical, and algebraic expressions (defined by the Content Language) that act or operate on instances of the ontology. The ACL content language defines three basic kinds of content-processing expressions:

1. *Logic propositions* that are evaluated to be true or false, e.g., used in the content of an 'inform' or a 'query-if' type CA
2. *Action expressions* to define single and aggregate actions, e.g., used in the content of 'request'-type CAs
3. *Identifying expressions* that identify an object and hence allows the same action to be looped through different object instances, e.g., used in the 'inform-ref' and 'query-ref' type CA.

Different knowledge exchanges will hence lead to the use of different combinations of CA and content expression being used, such as a sender *queries* a receiver to know *if data X exists or not,* vs. a sender *requests* the receiver to *act to send* it any data-matching constraints identified by X. Multiple Content languages (e.g., W3C-RDF, FIPA-SL, and constraint-based languages) may be used to support different kinds of processing expressions. FIPA-SL (FIPA Semantic Language) supports first-order predicate logic expressions. Although FIPA-SL is the main logical model used within many pure MAS applications, it is little supported outside the FIPA community. This is the case, e.g., when MAS applications are embedded as part of a more heterogeneous, non-agent, computational infrastructure.

There are pros and cons to consider when constructing a framework to express and reason about semantic structures. The conceptual model, and the logic and algebraic expressions as well as operators, can be either tightly or loosely coupled. The advantage for the tight coupling is that there is only one integrated representation and computation for both. However, it may not be possible to define a single set of processing expressions that is suitable in all application domains. For example, the OWL language (see Chapter 9) cannot currently express temporal and probabilistic expressions. Hence, it may be better to support multiple sets of different processing expressions as needed and a loose coupling between these. This is done in the FIPA ACL framework, e.g., to support probabilistic expressions in the weather-forecasting domain, or to support algebraic expressions in the book-accounting domain. In practice, the processing requirements of the application domain should determine the choice of content language and ontology language. For example, in the environment information retrieval the main processing requirements are SQL-type information retrieval requests. Hence, in this case a content language for EDEN-IW was chosen that encodes SQL-type actions, such as 'select', using an RDF content language.

Although the ACL model seems complex, it is a result of needing both processing flexibility and interaction flexibility. MAS software development methods, APIs, and toolkits can hide much of this complexity from application developers and users.

8.2.4 Communicative Act Semantics

Not only does the content part of the ACL message benefit from using a formal semantics to help organise and reason about the content (see above), but also the ACL header and in particular the communicative act protocol similarly benefits from having a formal semantics.

In fact, the two are related: the semantics of the content concepts depend, to some extent, upon an application context – acting when invoking the concept. For example, different values of determinants for water will have a different meaning for EU citizens depending on whether the water is used to drink, fish in, or swim in. Also, at a protocol level, the types of communicative acts link to specific supporting concept structures in the content. For example, a 'request' communicative act requires a secondary action specified in the 'content', such as 'register service' or 'update data' (Agent A requests that agent B updates data X). A second example is that a 'query-ref' communicative act requires an object-reference to be defined in the 'content', in order to contain the results of the query.

Thus, the use of concepts within a communication context (such as a query by a certain type of user with a certain intention) can, in part, determine the meaning of concepts. As mentioned previously, one set of semantics for communicative acts can be based upon modelling the intention of the sender. There are varying degrees of intention, from weak to strong that can be communicated from sender to receiver. There are pros and cons to agents being designed to support a stronger vs. a weaker notion of intention. The stronger intention notion imbues agents with a rich human-like mentalistic capability. However, this may be computationally intractable, or requires the evaluation of a huge state-space of possible intents. This state-space of intentions becomes even larger if the effect of the communication intention on third parties and society is also considered. For this reason, EDEN-IW did choose an intentional semantics for the communicative act model, which was based upon weaker intention of merely communicating. This was chosen to get a specific reaction in a specific receiver, rather than directly causing a change in the beliefs in the receiver.

8.2.5 Abstract Architecture Specification

The FIPA's specifications attempt to cover generalisations and high-level neutral abstractions. The core FIPA specifications are neutral with respect to specific services, end-user applications, software or hardware infrastructures, and implementations. The FIPA Abstract Architecture (FIPA001, 2005) defines a high-level organisational model for Agent Communication, and includes core communication support services such as directory, message transport, and agent communication services. The Abstract Architecture is neutral with respect to any particular directory service or the use of a particular network protocol for message transport. The Abstract Architecture itself cannot be directly implemented, but instead forms the basis for the development of concrete architectural specifications (Figure 8.3). Concrete implementations can implement all or part of the specification.

Specifications are often dependent on other horizontal layers, e.g., they may use an existing software infrastructure. A key issue is how much leverage to make from existing (non-agent) technologies and how to interface the agent parts to an existing infrastructure. The scope of MAS specifications generally includes the interpretation and handling of ACL messages, facilitator agents, and the use of existing software infrastructures, such as message transport protocols and message persistence schema, to underpin Agent Communication.

Different levels of granularities need to be modelled: we need to distinguish between agents and the level of granularity below that – the components that form the agent but which in isolation are not an agent. A related issue is determining which features should be modelled as agents and which should be modelled as non-agent parts. For example, the message transport specification was initially modelled as an agent in an early version of the message transport specification. At first sight, this seemed to offer great flexibility: because the transport agent is an agent, all agents can interact with it in a standard way and with a potentially very flexible and semantically rich ACL message interface. But there is a downside – efficiency. Transfer of a single message between agents always requires sending at least two messages, one to ask the transport agent to send a message, the other for the agent transport to actually send the message. Hence, in later specifications the message transport is specified as a non-agent service that can be invoked via some Application Programmer's Interface rather than via an ACL message interface.

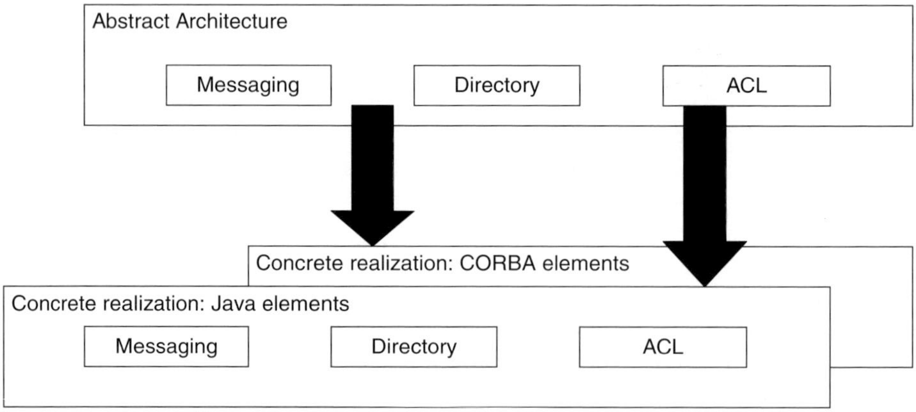

Fig. 8.3. The Abstract Architecture mapped to different instances of it.

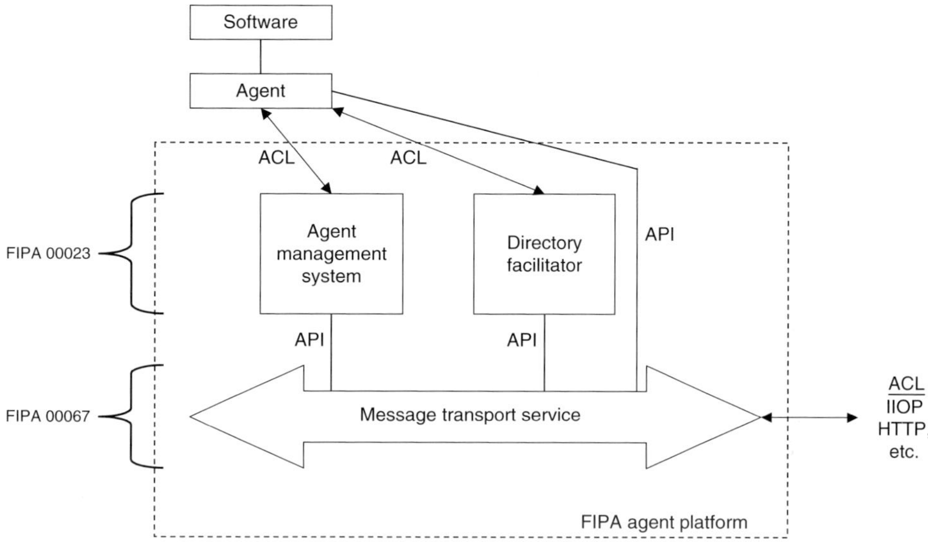

Fig. 8.4. The FIPA agent platform as defined in specification 23, the FIPA Agent Management Specification. The message transport service is defined in FIPA specification 67.

It is not mandatory that the services defined in the FIPA Abstract Architecture (FIPA023, 2005), such as naming or directory services, are agents, but they may be agents. In contrast, the FIPA Agent Management specification mandates agent name and agent life-cycle management service (called the Agent Management Service or AMS) (Figure 8.4) to be an agent. Agent-based name service, directory services, and the transport service can adhere to the properties and interfaces defined in the Abstract Architecture.

8.3 DEVELOPING MAS APPLICATIONS

8.3.1 Designing MAS Applications Using an AOSE Methodology

There are a variety of Agent-Oriented Software Engineering (AOSE) methodologies that can be used to develop MAS applications. The main types of AOSE methodologies (Iglesias et al., 1998; Wooldridge and Ciancarini, 2001) include: extensions to Object-Oriented Software Engineering (OOSE) methodologies such as AUML (Odell et al., 2001), knowledge-engineering methodologies, and formal methods. To illustrate the design of a FIPA agent application, a sub-set of an agent extension type of the OOSE methodology will be used. Furthermore, some core design models for different design viewpoints will be derived, and the implementation of this application using FIPA agent toolkits will be discussed.

In order to keep things simple, all agents are considered to behave reactively (i.e., to have weak mental intentions). These agents use simple protocol-based semantics for the communicative acts, i.e., they send messages in order for the sender to get a specific response from the receiver (Ferber, 1999). The key design models or viewpoints considered here are tasks, roles, and interactions. The task model defines the tasks hierarchy that has to be undertaken in order to achieve a goal. The roles define a group of functional parts that an agent plays in a joint process within an organisation. The interaction model defines a pattern of messages exchanged by two or more agents in a joint process such as problem solving.

As an illustrative example (Figure 8.5), the EDEN-IW task agent is triggered to help a user agent to locate data resources, and to find alternate data resources that are no longer in a known location. There are three interaction options to locate the assistant: query for assistant in a known location, query the store directory, or perform a broadcast. Query, inform, cfp, and propose present FIPA communicative acts (see Table 8.2).

8.3.2 Task Model

There are many types of task models. The one applied here is commonly used in the field of Human Computer Interaction (HCI), which is called Hierarchical Task Analysis (HTA) (Annett, 1967). The main characteristics of HTA are:

- Goals are decomposed into sub-goals. (In HCI it is based on a person's mental representation of the activity.)
- Each sub-goal and activity may be described by verbs and names.
- Each sub-goal or activity has a number.
- The plan at each level of decomposition and on each branch of the tree indicates how the activities are done (sequential, parallel, or simultaneous) and their characteristics (optional, conditional, or loop).
- Each plan has a number.

The decomposition stops when it is not possible to break down the cognitive activity any more by the person, or when the analyst considers that more decomposition would not be useful for the application.

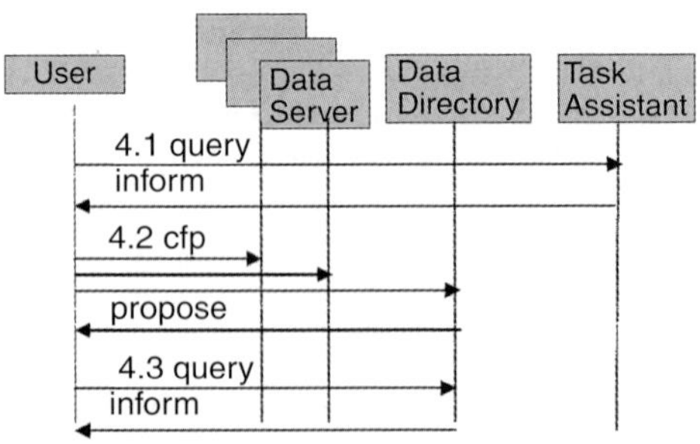

Fig. 8.5. An interaction model for a part of a task model that deals with locating help in the form of an assistant. The number refer to the tasks defined in Table 8.2.

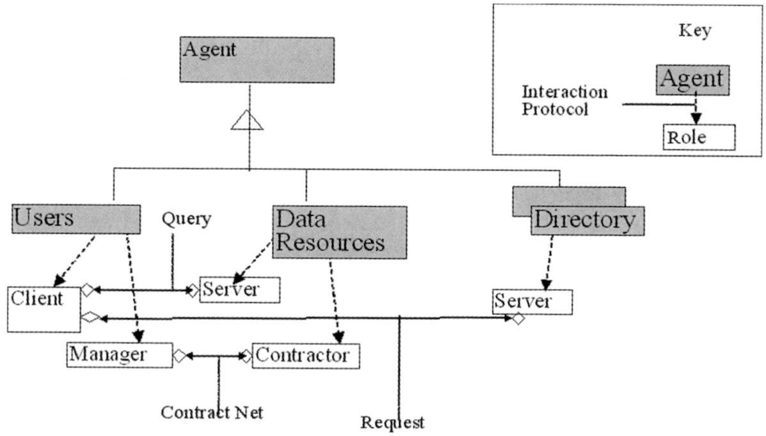

Fig. 8.6. Role model for the locate task assistant sub-task (Task 4, Table 8.2). This model is expressed using an UML object diagram.

A HTA for a help system to locate a resource is given in Figure 8.6. There is considerable flexibility in the design for locating a resource that is no longer in a known location. The user agent may rely on internal knowledge, may search for assistant, or may try another strategy. For each of these options, there are sub-options.

8.3.3 Interaction Model

It is possible to map the task model to an interaction model by designing tasks as combinations of message transmissions and message handlers. Message transmission can be modelled using different levels of intent (Ferber, 1999). For example, a broadcast request message can be sent to anyone to retrieve item X at location Y, or to request help from a specific receiver such as a task assistant, a directory service, or a data service. The interaction model in Figure 8.5 uses a simplified form of a Unified Modelling Language (UML) sequence diagram that omits task duration and task concurrency. The Contact Net interaction in the interaction model has been over-simplified into an exchange of two messages: a cfp message and a proposal response message ('propose'). The interaction model identifies the agents at the start and end points of the interactions.

8.3.4 Role Model

The third viewpoint of our design represents a type of organisational or structural model. A role can be defined as a logical group of atomic activities such as interactions according to the application context. Roles can be modelled using a UML object diagram (Figure 8.6).

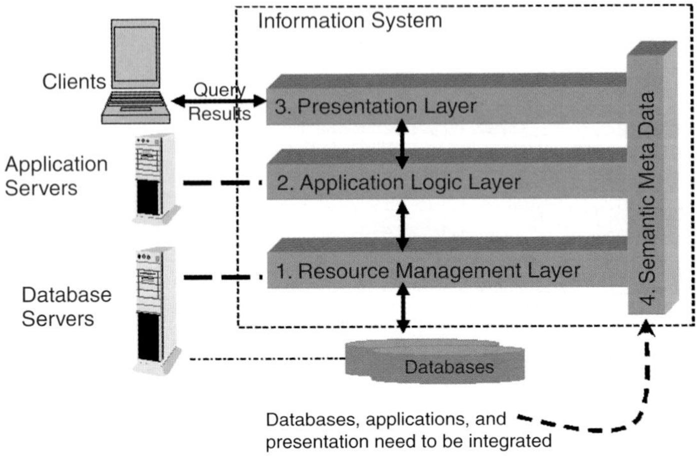

Fig. 8.7. Standard model of an information system.

In the role model, agents are modelled as dependent elements (shown shaded in Figure 8.6). They depend on the roles modelled as independent elements. The roles in turn are linked to the interactions. For example, the FIPA contract–net interaction defines contractor and manager roles. Hence roles in different agents are associated with each other via particular types of interaction: an interaction is modelled as a named association between two or more agent roles in the diagram.

8.3.5 FIPA Agent Toolkits and APIs

There is inherent high complexity, cost, and risk in developing actual agent applications. This acts as a barrier to users and developers who wish to assess this new technology. Clearly, reference implementations are useful. Open source toolkits of the FIPA agent specifications such as FIPA-OS, JADE, and ZEUS (Poslad et al., 2000) can reduce the barrier for the adoption of FIPA standards, enhancing the ability of agent application developers to construct applications using FIPA technology. FIPA toolkits provide the following useful features:

- agent shells and templates with inbuilt support to send and receive FIPA ACL specification messages, and to access the standard FIPA agent services for Agent Management and directories
- codecs and decodecs for ACL messages to access and construct the ACL message fields such as the content, interaction protocols, and sender and receiver IDs
- support for managing and storing messages and for scheduling the standard FIPA message interaction protocols such as query, request, and contract-net
- tools for monitoring and testing agents.

In the implementation phase, these design models can be refined to map to the various elements, messages, protocols, and services defined in the FIPA specifications. For example, the role organisation model can be mapped to the agent platform model defined in the FIPA Agent Management Specification. Many FIPA toolkits support some kinds of task objects, including 'receive ACL message' tasks, 'send ACL message' tasks, and 'receive then send ACL message' tasks. So the design task model can be mapped onto FIPA toolkit task objects. At one level of abstraction, the design of a reactive FIPA agent corresponds to the definition of a set of ACL 'receive message' and 'send message' tasks.

Toolkits provide support for some of the basic message exchanges. For example, toolkits will provide preset tasks for AMS registration, queries, and AMS de-registration. Toolkits also provide support for task scheduling. If an agent toolkit is implemented using a procedural or object-oriented language, very basic plans can be expressed as hard-coded sequences of task calls interspersed with conditional statements to evaluate the success of previous tasks.

8.4 EDEN-IW MAS SYSTEM DESIGN AND IMPLEMENTATION

The conceptual architecture of EDEN-IW information system follows a conventional three-tiered information architecture design (Figure 8.7) consisting of a resource management layer, an application logic layer, and a presentation layer. In a heterogeneous distributed system, such as EDEN-IW (agent) functional components in each of these layers can be distributed and heterogeneous. In the EDEN-IW system, functions in each of these layers are integrated using a Semantic Web metadata model integrated using a multi-agent infrastructure.

Each of the main three layers, such as the user portal presentation layer, may be so complex that it, in itself, is internally organised as a tiered sub-architecture. A multi-agent system is a good potential architecture for integrating heterogeneous databases because agents are naturally distributed and autonomous; they can use rich explicit communication protocols to interoperate and they can naturally link to semantic models to help resolve interoperability problems.

Each of the agents has a specific task in the complex process leading from the question raised by the user, through the formulation of specific (but database-independent) queries, through to the specific queries sent to databases which the agents evaluate as potentially having an answer or part of the answer for the actual question. Such requests for information require a common 'language', a list of accepted and well-defined words, which is the basis for an ontology-based Semantic metadata model relevant for inland water. When results are returned, post-processing will be performed in order to furnish the user with information in the most useful form. This post-processing consists of harmonising, aggregating, and presenting information in a consistent form, allowing variation in the level of detail presented, and integrating decision support tools for environmental management for the benefit of policy makers.

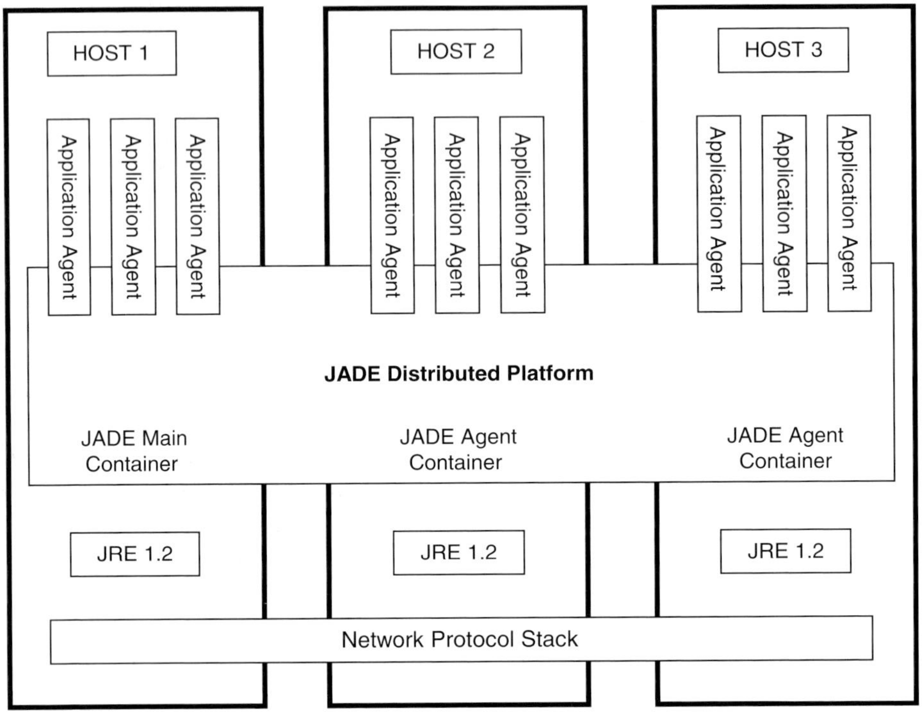

Fig. 8.10. JADE agent technology view of the EDEN-IW system.

plans are made according to different queries, e.g., one or more plans for different use cases such as '*What is the concentration of X in River Y at time T?*' or '*Which stations have data on Determinand X?*'

- *Directory Agent* (*DA*) is the central repository for metadata. It stores a list of all available stations and active resource agents. It also holds a list of all determinands available at a particular database. These resources are monitored and this information is regularly updated. The Directory Agent helps the task agent to locate the appropriate resources for each use case. At a lower level, it keeps a list of all active agents in the EDEN-IW system. Agents can register, deregister, or search the Directory Agent for other agents
- *Resource Agent* (*RA*) provides the resource-wrapping service to access conventional databases to retrieve the IW data. A Web access model has been built to secure access to the database information. The wrapping service accesses both EDEN Global View (EGV) and Local Database View (LDV) ontologies via appropriate ontology 'adaptation' services. The heterogeneity of the inherent database structure and language representation has been hidden from the common EGV language and semantic

representations. Thus, the Resource Agent can translate the global query from EGV to LDV and the local database query result from LDV to EGV in reverse. In the prototype demonstration, two resource agents have been created, IOW and NERI: each agent accesses the corresponding LDV in the ontology storage for EGV/LDV translations

The EDEN-IW system is implemented as an open source Java agent platform called the Java Agent Development Environment, or JADE (Bellifemine et al., 2001), and a set of domain-specific EDEN-IW application agents and non-agent software services, see (see Figure 8.10).

The JADE Agent platform provides the following core agent service: Agent Communication and Message Transport, Agent Name Service, Agent Directory Service, and Agent Management Services, monitoring and configuration tools. To traditional software engineers, the agent technology appears as a set of distributed Java applications that are interlinked using a combination of Java RMI and XML messaging over an agent transport such as HTTP.

The EDEN-IW Java application agents run within the JADE distributed agent platform infrastructure. The agents exchange IW metadata in a common form called the EDEN Global terms or the EDEN Global View (EGV) terms – this insulates the majority of the agents and the user from needing to be familiar with the local database terms.

The EDEN-IW System followed an evolutionary life cycle of development in order to engage potential EDEN-IW system users as early as possible. During the project, a number of prototypes were developed with increasing complexity.

8.4.1 Agent Message Interfaces

The content of the Agent Messages in the EDEN-IW system is represented in RDF rather than in DAML+OIL. The main reason for this is that RDF, but not DAML+OIL, has been proposed as a FIPA content language. Furthermore, service action invocation does not need very expressive semantics. Instead, it often needs to be simple and quick to invoke.

The ACL message headers provide a generic communication context for service-specific information and actions, e.g. the ACL message defines a message type (communicative act) such as query, and the ACL message body defines the parameters or constraints for the query. As an example, consider the user interface that collects and validates parameters that have been entered by the user. The corresponding User Agent then creates an RDF message to interface the rest of the agents in the EDEN-IW System. This message body contains all the information required by the agents. It specifies the following FIPA Tags:

- *Action Tag*: This provides unique message identification.
- *Actor Tag*: This defines the actor (agent) who initiated this action.
- *Act Tag*: This defines the action to be carried out e.g. 'GetParameter'.
- *Argument Tag*: This gives the input parameters required to perform the query. These constraints are specified using the Global EDEN ontology.

- *Done Tag*: This is used to track the status of the process being performed, such as 'start', then 'pending', and finally 'done'.

The FIPA ACL messages can be encoded using XML as defined in FIPA specification 00071 (FIPA071, 2005) (Table 8.3). When an agent wants to send an XML encoded message, the API used by the FIPA platform is called. The message to be sent is then created by the API, and the fields of the message are filled in. Agent communication messages are transported over HTTP and Java RMI (Remote Method Invocation). Agents located on the same machine use RMI to call methods on each other. Communication with agents on other machines is done over HTTP. Thus, an additional transport envelope header in XML, as specified in FIPA specification 00085 (FIPA085, 2005), is added, when agents are sent by FIPA agent transport services in different agent platforms. Agents do, however, not see this transport envelope header, as it is stripped away by the message transport service.

8.4.2 The User Agent

The User Agent collects parameters that have been entered by the user at a Web-based user interface. Parameters are validated by the user interface. Then an RDF/XML message is generated from these parameters as described above to enable other agents in the EDEN-IW System to cooperate to solve the task of answering the user query. The RDF part of the message is the content of the ACL/XML header message that is passed from the User Agent to a Task Agent.

8.4.3 Agent Tasks and the Task (Planning) Agent

Each agent supports a series of tasks. Tasks are goal-orientated and application-specific service operations. The core agent tasks are: messaging, agent management, 'hello', and 'ping'. Agent messaging is the core task to send and receive ACL messages. Agent management tasks control the life cycle of the agents. A 'hello task' allows other agents to ask an agent directly, rather than via the directory service, about its communication and service capabilities. Ping tasks respond to 'ping' messages to confirm that an agent is alive. In addition, there are agent-specific tasks: the Directory Agent supports tasks to retrieve metadata information, e.g., a query to find out which agent(s) hold information about a particular river or the time range a database spans.

The tasks of multiple agents often need to be combined and coordinated because no single agent has all of the information and capabilities to perform a task itself. A special agent, called the Task (Planner) Agent (TA), is used to plan and coordinate the composite tasks of multiple agents. The composite tasks include the following:

- *Data fusion tasks*: Upon receiving several results from multiple (resource) agents, one of the most complex tasks is to merge the results together. Often results are in different units, so a conversion is required before displaying them to the user.

- *Spatial tasks*: These tasks are involved in gathering of data within a specific geographical area. The first step is to locate all the rivers, stations, or basins, depending on the query. It sometimes also involves gathering data from several data resources.
- *Temporal tasks*: These tasks involve the gathering of data that spans a particular portion of time. Data sets range over long periods and user queries may be limited to a specific time period. The user may also ask to be informed whenever a particular event, related to a measurement criterion, occurs. The Task Agent coordinates the interaction between the User Agent, Resource Agent, Directory Agent, and other agents.
- *Decompose query tasks*: These tasks decompose queries into answerable ones, for the Resource Agent(s) and Directory Agents to answer.
- *Query status reports*: These tasks audit the queries processed in Task Agent.
- *Exception handling*: These tasks focus on reporting errors to Web page in user terms, e.g. when the results from the database are not exactly the ones expected by the end user. Alternatives can be offered to users such as searching for other relevant information.

To coordinate the interactions between the different Agents, the Task Agent plays a role in harmonising communications. According to its functional requirements, the Task Agent consists of four major parts: decomposition unit, assembly unit, exception handling unit, and task scheduler unit. The Task Agent knows the roles of other agents, i.e. the functionality, and the kinds of problems they are able to solve. The task description is updated and maintained by the Directory Agent, and is provided to the Task Agent. The decomposition unit breaks down the incoming queries into sub-queries that can be handled in the ontology service and by the Resource Agents. It then allocates some actions accordingly to each sub-query. The scheduler unit schedules the actions, while the exception-handling unit monitors and deals with any exceptions that may happen. Finally, the results from the Resource Agents are assembled in the assembly unit and sent back as replies to the User Agent.

According to the task description, each task is one step towards an agent's goal. In the case of querying inland water databases, answering a query is a goal, which is decomposed into tasks and sub-tasks according to a plan. For example, the task decomposition of the query 'Which sub-river basin should be cleaned first?' is listed below:

- Extract information about the concept 'sub-river basin' from the ontology service.
- Query the Resource Agent (RA) and database (DB) to find out the relevant stations that form a sub-river basin.
- For each sub-river basin, send a query to a specific RA and DB for specific data about the water quality, and collect the responses.
- Perform analysis and post-processing.

A plan is an organisation of a series of tasks in the way that tasks and results are handled efficiently. The production of a plan is driven by different use cases and is based on specific rules. The FIPA ACL protocol that EDEN-IW agents use to interoperate naturally supports diagnostics. There are three standard ACL message types corresponding to error handling:

1. *Not-understood*: The agent tells another agent that one or more parts of a message are not understood.

2. *Failure*: The action of telling another agent that an action was attempted, but the attempt failed, e.g. the agent did not have access permissions to use a resource.
3. *Refuse*: The action of refusing to perform a given action and explaining the reason for the refusal, e.g. too busy.

The Task/Planner Agent detects these error messages. Depending on the application context, the error may be passed to the User Agent to notify the user or handled internally by the system. All the agents in the EDEN-IW system support the 'ping' interaction. This can be used to check whether agents are responding or not. In addition, the JADE agent platform itself has tools to support interaction visualisation and message audits (Sniffer agent) and to test message generation (Dummy agent).

The TA is currently implemented as a JADE agent and simple plans are implemented using Java language constructs. As far as scalability of the system is concerned, there are two ways of handling this. A pool of (simple) task agents is available, or the task agent is designed multi-threaded to handle a greater number of incoming queries.

8.4.4 The Directory Agent

The EDEN-IW Directory Agent acts as a repository for Agent Descriptions and Service Descriptions. Agent descriptions define the Agent-name, Agent-locator (One or more transport-descriptions), the tasks they support, the domain Ontologies, and the Interaction Protocols they support (FIPA, 2005a,b). The service descriptions for the resource service entries in the directory are expressed in terms in the EGV Ontology such as the key concepts of geographical regions (e.g. groups of stations), Inland Water parameters, and time.

The directory agent compiles summary metadata about each database resource. For example, it knows the determinands, the time range, and the list of stations that a database covers. When a service agent such as a resource agent is started, it can advertise itself in the directory agent, and thus makes itself available for use (client data pull). In addition, user agents can register their preferences, and the directory agent will periodically contact the data resource providers for updates and then notify the user agents when new service capabilities come online to match user preferences (client data push). The latter interaction is provided through the FIPA specification: Subscribe Interaction Protocol (FIPA, 2005). The directory agent supports the registration, de-registration, and modification of registrations of agents and services. It also performs agent and service matches when queried. In addition to the above-mentioned basic functions, the directory agent can periodically check for the existence of registered agents and update its internal directory about their status.

8.4.5 Ontology Services and the Resource Agent

The EDEN-IW system uses a multi-agent system to process and distribute the EDEN-IW semantic metadata models. In order for agents to function as metadata processors, they must import Semantic data messages using an HTTP transport, then parse, and verify them in order to process them. The EDEN-IW agents use HTTP to import Ontology

Table 8.2. Task model (HTA) for a goal to retrieve item X, such as data matching a query constrain, at location Y such as in a database and the subsequent use of a task assistant to help plan what to do if this does not succeed. One of the plans, plan 0, for organising part of the task hierarchy is also shown

Task 0: Retrieve item X.

- Plan 0 [Do 1 then 2; If result of 2 = yes, then do 3; If result of 2 = no, then do 4; If result of 4 = yes, then do 5; If result of 5 = yes, then do 6; If result of 6 = no, then do 7]
- 1: Retrieve Item X at known location Y
- 2: Check if Item X at location Y
- 3: Retrieve Item X at location Y
- 4: Locate Task assistant
  - Plan 4:
  - 4.1: Search for assistant in known location A
  - 4.2: Broadcast if anyone can help X
  - 4.3: Query Directory Service for assistant
- 5: Check if assistant available
- 6: Ask Assistant about Item X
  - Plan 6:
  - 6.1: Ask Assistant to retrieve Item X (delegation)
  - 6.2: Ask Assistant to locate Item X
  - 6.3: Retrieve item from new location
- 7: Modify Search for Item X
  - Plan 7:
  - 7.1: Check for substitute item for X in Location Y
  - 7.2: Search for Item X in location Z
  - 7.3: Wait for item to become available in location Y

Table 8.3. A fragment of a FIPA-ACL header in XML

```
<?XML version ... >
xmlns="http://www.fipa.org/schemas/acl#">
<communicative-act> ......
....
<content> ...</content>
```

documents from known locations. The Jena Semantic Web Java toolkit (McBride, 2002) is used to parse DAML + OIL messages to extract data of interest. Generally validation is not performed.

The mapping between different parts of the Ontology model that need to be related and between the Ontology model and the non-ontological external database resource instances is not always simple. For example, the same data concepts may be structured quite differently in different databases.

It is sometimes not sufficient just to use XML namespaces to link terms between different ontologies. Some conversion or the use of some formulae may often be necessary to link terms. For example, when converting one set of measurement units to another set of measurement units. For this reason, EDEN-IW implements Ontology mapping services, e.g. the EGV to LDV-IOW (Local Database View of the IOW database) mapping service. On start-up of the Resource Agent, it reads in and parses the global and local ontologies. It is then able to translate terms between instances of the EGV and LDV-IOW ontologies. This ontology mapping uses graph theory, and it was implemented using Java in the first version.

Introducing a new database resource
There is a well-defined process to introduce a new IW database resource into theEDEN-IW system. It consists of two major steps:

Semantic data model instantiation
1. Define a Local DB View (LDV) of the IW database structures represented as a DAML + OIL document. This is done in cooperation between the database owner and EDEN-IW domain expert. It is often generated from out-of-date documentation of the DB model.
2. Map the LDV to the core concept model (global view – EGV).
3. Map the LDV to the local Database resource (SQL queries).

Semantic data processing
1. Configure a new resource agent to wrap the new resource to import and to parse the LDV ontology instance.
2. Start the new Resource Agent as part of an existing (or a new) agent platform. The Resource Agent then automatically registers metadata summaries of its stored data in a directory service agent and makes it available to the whole EDEN-IW system. This involves the database owner implementing the existing LDVs as examples or templates.

8.5 DISCUSSION AND CONCLUSIONS

The FIPA framework has proved to be a powerful and flexible approach that can integrate semantic Web messaging with a flexible rich set of communication protocols to enhance information retrieval. Pragmatic choices are needed when MAS need to be embedded as part of heterogeneous infrastructure that uses XML message exchange and several design choices reflected this. For example, the FIPA-RDF rather than FIPA-SL has been used as the content language to define the content computation in a declarative manner, DAML + OIL was used as the Ontology language and XML was used for the message header and body encoding. Thus, from an implementation point of view, the EDEN-IW system appears as an XML-based message exchange system.

Only a few of the rich set of possible FIPA interaction protocols have been used, mainly the FIPA-Query interaction protocol and FIPA-Request for some of the directory service management tasks. This had more to do with the necessity of simplifying the system in order to be able to develop it in the time frame of the project, rather than a comment that the richness of interaction in the FIPA model is not useful in practice. EDEN-IW adopted semantics for Agent Communication that was based upon a weak intention reactive model, senders communicating in order to receive a specific reaction from the receiver rather than senders intending to directly change the belief state of the receiver.

The EDEN-IW MAS is open sourced. It has been based upon the JADE agent middleware, and extends JADE in several ways.

The EDEN-IW agents can parse the message content expressed in RDF and the DAML + OIL ontology. The RDF language is used to express the triplets of service actions, the actor that calls them, and their constraints. The DAML + OIL language is used to express the relationships between the different schema instances for the databases, against a global database schema.

The EDEN-IW uses a wrapper model inside its Resource Agents (RA) to generically wrap relational database resources, and to make their information accessible to other agents in the system. EDEN-IW has a task-planner agent to coordinate tasks that involve the use of multiple agent services such as multiple data resources. A Directory Agent (DA) extends the JADE/FIPA Directory Facilitator (DF) agent to support the advertisement of rich service descriptions of the services on offer. In addition, the DA is proactive and uses additional standard agent interactions to enable it to be automatically updated when the status of the RAs or their resource descriptions change.

The JADE version 2.8, which was used by the EDEN-IW system, seemed to lead to occasional HTTP message exchange problems. It was known that the more recent JADE version 3 had an improved HTTP transport implementation (plus other useful features such as a distributed log facility). However, the conversion of the agents to be able to use the JADE version 3 API was not straightforward. Given the time pressure in the project, EDEN-IW continued to use version 2.8.

The EDEN-IW has demonstrated several important contributions to applying the MAS model. EDEN-IW demonstrates that agents can promote an integrated virtual information system comprising users, processing, networks, and information resources. It demonstrates the use of agents as semantic message processors to reason about relationships between terms in different local semantic information instances and the global semantic information model. It shows how agents can act as brokers and can cooperate to simplify the interaction between multiple service providers such as when an information query requires results from several heterogeneous database resources and when any heterogeneous results need to be harmonised.

Chapter 9

AN ONTOLOGY-BASED APPROACH FOR ENHANCING INLAND WATER INFORMATION RETRIEVAL FROM HETEROGENEOUS DATABASES

M. Stjernholm, S. Poslad, L. Zuo, O. Sortkjær and X. Huang

9.1 INTRODUCTION

A semantic mediation approach is used to integrate distributed inland water database resources and heterogeneous types of users and their applications. In the EDEN-IW approach, the core technology is based upon a combination of an ontological model to capture semantic metadata (Fensel and Brodie, 2003) for inland water, software agents as an infrastructure to share and reason about the inland water semantic data model (Poslad and Willmott, 2003) and XML to make the information accessible via Web portals and XML-based Web services.

The EDEN-IW project focused its development efforts within the restricted domain of surface water and more specifically on stream water data, although the concepts behind EDEN-IW may be applied to other domains of water and to the environment.

This chapter describes the structure of a layered Ontology model and discusses practical issues related to the development and maintenance of the Ontology.

9.2 WHAT IS AN ONTOLOGY?

Ontologies are conceptual models that can be used for knowledge sharing. An Ontology is characterised by: Being an explicit semantic model of the concepts and structures used to represent and manage them, representing information that is machine-readable, and by the consensual nature in agreeing and sharing this model. More exactly, Sowa defines Ontology in the following way: "The subject of Ontology is the study of the categories of things that exist or may exist in some domain. The product of such a study, called an Ontology, is a catalogue of the types of things that are assumed to exist in a domain of interest D from the perspective of a person who uses a language L for the purpose of talking about D. The combination of logic with an Ontology provides a language that can express relationships about the entities in the domain of interest".

not available online nor standardised. Metadata concepts are often better representations of the world if they are based on hierarchical models such as object-oriented model classes with attributes, subclasses, and relationships to other classes rather than on flat table relationships.

The Danish database (NERI) and the French (IOW) are two examples of heterogeneous databases that were initially used. They have been used in the development of the EDEN-IW Ontology model.

The NERI inland water database is structured as a number of observation programs, where each program has its own set of tables. The observation programs cover both research projects and monitoring programs without public access.

The objective of the IOW database is to distribute data, usually as products after processing. The stored data comes from national thematic databanks and the River basin data banks. The technical architecture is based on an ORACLE database server, and an ARC/INFO server for map processing.

In addition to the structure of the databases, other differences between NERI and IOW were the understanding of simple concepts such as the medium "water". A water sample from a lake or river includes small organic or inorganic particles (and even fish), which can be filtered and divided into a water and particle phase. A determinand like Nitrogen can be found in the water fraction as well as in the particle fraction, or it can be analysed as total nitrogen in the water sample. It is obviously very important to define every determinand precisely to ensure that at least the main concepts are commonly accepted, as the concepts may not be represented in the same way in different databases. If the definitions of the determinands in the relevant databases do not correspond to each other, new intermediate determinands may be needed.

Although the main concepts may be commonly accepted, local implementations can vary substantially. Similar observations may be handled differently in different database implementations. The result of the differences between databases, see Table 9.1, is that data from different databases requires additional concepts and primitive classes to be defined.

Table 9.1. Different implementations of observations in a French (IOW) and a Danish (NERI) database

Database 1 (IOW)	Database 2 (NERI)
• Each Observation value is linked to a Determinand and an Analytical fraction (local codes). • Each combination of Determinand and Analytical fraction is linked to a specific Unit defined in a Data dictionary (text document). • The Analytical fraction is implicitly linked to a Medium.	• Each Observation value is linked to a Determinand (local code). • The local Determinand name (in Danish) implies the Medium and Analytical fraction. • Each local Determinand is linked to a specific Unit (local code).

9.4 EDEN-IW SEMANTIC MODEL AND SERVICES

Developing and deploying a semantic approach to integrate data resources and users is a very powerful approach. However, it faces several key challenges such as: the semantic model needs to be defined; it needs to be mapped to local database models; it needs to be logically and physically accessible via a variety of different viewpoints according to different application and user requirements. A life-cycle model is also useful to allow the semantic metadata model to adjust to meet changing user requirements.

The semantic metadata model contains as the core design a separation of the conceptual model from (a) the application, (b) the user's commitment to use the conceptual model, and (c) the local database models. This has the advantage of making the conceptual model more reusable and maintainable across applications and heterogeneous databases within a domain.

The core conceptual model, called the EDEN-IW global view Ontology (EGV), defines all the concepts necessary for describing the world of inland water quality. In addition, each database resource, user, and application defines a separate data model that is mapped to the global model. Thus, each of the connected databases defines a local database view Ontology (LDV) mapped to the EGV model. The classes and facets of the LDV Ontology will be subclasses and aggregations of the EGV set of classes and properties. Ontology services perform the knowledge mediation between the EGV and the LDV models, and between the EGV and other data models such as multi-lingual data dictionaries and standard glossaries of terms and user preferences (such as their preferred constraints for queries and metadata queries).

The EGV Ontology model serves several purposes:

- It provides a common data dictionary – definitions, concept names, and enumerations of, e.g., determinands and units.
- It provides the basic classes (primitives) for creating local concept classes.
- It provides a schema of the information required for each concept, e.g. an "observation" requires more than just a value and a unit to describe the type and context of the observation.
- It provides an organisation of common knowledge about relationships among the classes.

Analysis of the domain of "Inland Water quality" has shown that similar terms are used in the description of monitoring programs and observations. Deeper analysis has also shown that the understanding and implementation of the same concepts do differ in crucial areas, and can lead to misconceptions if they are not handled in a strict way. A common problem is that key information may not be expressed explicitly, but remains as "implicit common knowledge" within a local group of data managers. Documentation (of the metadata) often resides in a non-computational form, in printed documentation separated from the databases.

For practical purposes, the EGV Ontology may be divided into a number of separate ontologies. At the start of the EDEN-IW project, ontologies for Inland Water Quality did not exist. Therefore an Ontology had to be created from scratch. A common Ontology was created, based upon common concepts extracted and merged from non-machine-readable forms of metadata, such as unstructured documentation of different data sources. From these, common concepts such as "Station", "Determinand", and "Time", and common measurement concepts such as "Units" were identified. Concepts which relate to databases (e.g. "Tables", "Field", and "Primary Key"), independent of the application domain, were also included.

The architecture of the EDEN-IW data model (see Figure 9.1) is related to the three schema ANSI/SPARC architecture (Tsichritzis and Klug, 1978) that has a lower layer reflecting the local physical representation in the database, a middle conceptual schema, and an upper external schema that provides different view of the conceptual schema from the perspective of applications. In this context it should be noted that many network and information system models further refine the upper application layer into a processing layer and a presentation layer, so, for example, the same processed information may be presented in French and English.

The main advantage of this division into three basic layers is that it allows additional database models to be added without changing the other layers, providing they do not require changes into the conceptual model. Furthermore, new application uses of the conceptual data model can be added without requiring changes to the other data models.

The architecture in Figure 9.1 represents a schematic model. In practice, the different (meta)data models are contained in separate data structures which are expressed in representations such as the W3C OWL (Ontology Web Language). The software needs more than an ability to parse these different data models; it must also support the various

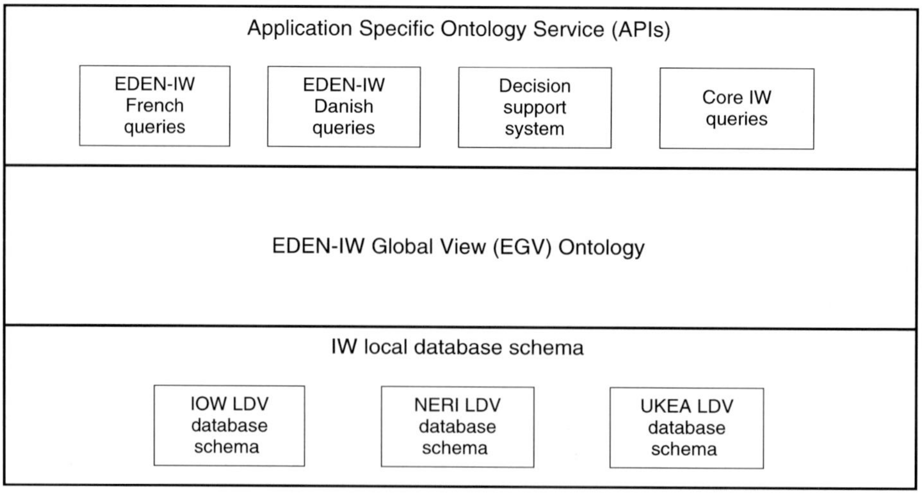

Fig. 9.1. EDEN-IW layered ontological model. The middle layer is the core concept model, the upper layer represents how services hook into the Ontology. The lower layer represents the local database models.

mapping functions. These include the transformations required to move to and from the local database model to the common conceptual model, and to the different application and user viewpoints.

9.4.1 Examples of Ontology Queries for EDEN-IW

Some examples of conceptual Ontology queries for EDEN-IW:

- Find the list of all concepts in the EDEN-IW Global View.
- Find the value list of property DeterminandID in the class Determinand of EDEN-IW Global View.
- What is the Determinand Name of DeterminandID 35?
- What is the mapping concept of Determinand in NERI domain?
- Find the value list of all relevant instances in the NERI domain related to Medium water.

Some examples of application-based Ontology questions:

- What is the column name for Determinand concentrations in stream samples in the NERI database?
- Find the SQL query to the NERI database for "What is the concentration of Determinand X in River Y in Time Period Z?"

9.4.2 Representation of the Ontology

Computer processing and machine deduction require that all knowledge about the data is specified and stored in a computer-readable way. When the project was proposed, and when the project application requirements were gathered, a proposal was made for an XML structure to document metadata and database schemas for the resource databases. A part of the structure was the use of so-called Field Keywords. However, the use of standard XML has two serious weak points:

1. The language structure is not sufficiently rich to express all the relationships and data type properties
2. No standard tools exist to extract and deduce knowledge from a knowledge base expressed in standard XML.

In the early 2000s, commonly used Ontology representations included DAML+OIL (DARPA Agent Mark-up Language + Ontology Inference Layer) and Knowledge Interchange Format (KIF). Of these, only DAML+OIL is based on XML. During the project period, another XML-based Ontology called the Web Ontology Language (OWL) has been proposed (Smith et al., 2002). Nevertheless, it was decided to maintain the core Ontology elements in DAML+OIL, since OWL was (at the time) not found sufficiently mature in terms of tool support. However, later on in the project, OWL was used for the representation of elements like UNITS and DETERMINANDS (see below).

Elements in the application commitment layer are represented as separate DAML+OIL documents, and are available on the EDEN-IW Web server. They are linked to the bottom layer, see Figure 9.1, which itself is represented as another DAML+OIL document, using XML name-spaces. The documents are separated into two parts, where the class specifications are in DAML format and instantiations of classes are stored as RDF (Resource Descriptor Framework) documents.

Applications such as the query applications with user interfaces define service APIs as sets of actions specified in RDF-S (Resource Descriptor Framework Schema). For example, the 'getParameter' action is defined as a part of the User Agent service interface. This action takes attributes that are concepts in the DAML+OIL global Ontology – these are again linked using XML name-space.

In theory, the system would have been simpler to develop if a single representation would have been used for the metadata and data models, i.e. for the common or global concept model, for the data sources or local database data, and for the applications' use of the data. However, in practice, multiple representations were used. For example, EDEN-IW made use of XML, OWL, DAML+OIL, RDF, RDF-S, and SQL for the metadata models. XML was used to encode and exchange the metadata using the FIPA agent communication protocols. DAML+OIL (and later OWL) were used to capture the application domain structure and (class) relationships. RDF was used for the instance of these classes. RDF-S was used to model the service actions or operations that act on the metadata and data, as it was a proposed standard FIPA agent communication "content language".

At the time when EDEN-IW was started in 2001, DAML+OIL was not mature, and was not a standard model for semantic metadata. Tools support like the HP Jena toolkit documentation advised that classes or data types should be specified in DAML+OIL. It also recommended that data instances should be represented in RDF in order for the application code to be able to access semantic metadata models.

Often, real world and legacy data sources such as relational database data are not in a suitable semantic metadata form. They need to be wrapped, mapped, or restructured to relate to the semantic representation. In addition, the metadata may need additional encodings and headers in order to distribute and to communicate about the metadata. Services that use the metadata to define properties of the service actions may also use a weaker Ontological form that is efficient to parse rather than expressive in order to support high throughput transaction performance.

Not only do metadata model developers need to develop skills in learning and understand new complex metadata representations, they must need also to appreciate that the use of semantic metadata is still evolving. This may introduce a drive to re-engineer metadata models from one representation into another (newer one), to assess how equivalent these are and to deal with how to relate heterogeneous metadata representations during development.

9.4.3 Global View Ontology

In order to encompass a variety of local database implementations exemplified in Table 9.1, the EGV is largely made up of "primitive" classes. The current EGV includes classes that are specific to the Inland water domain, as well as more universal classes. The latter are suited for describing database schemas and elements such as Time and Units. The classes are organised in hierarchies with *EdenGlobalConcept* as a superclass. The EGV also contains relevant instances of the defined classes.

The Inland water databases contain information of the type *the VALUE of DETERMINAND observed at a STATION at a TIME.* A deeper analysis of the concept of "the VALUE of DETERMINAND" in a couple of databases has identified that the value of a determinand may actually express different types of information:

- Instantaneous values vs. time-aggregated values.
- The same determinand may be observed in different media and fractions of the medium.
- The value may be expressed with different units, e.g. milligram/litre or nanogram/litre.
- The value may be expressed in different chemical compounds, e.g. concentration of Nitrate may be expressed either in milligram N per litre or in milligram NO_3 per litre.

Hence, this has led to a model of global class relations for determinands that supports these design requirements (see Figure 9.2).

9.4.4 *Class vs. Instance Representation Issues*

The use of any representation of a data model necessitates conforming to restrictions of the constraints in expressivity of that particular data representation. When application users and application domain experts start to develop a domain model, the approach is often to examine instances of classes and relationships between instances, i.e. the concrete instead of the abstract data. There may be a desire to capture relationships between instances rather than to view this more abstractly as relationships and constraints on classes. Some instance constraints can be captured in terms of specifying classes whose properties have certain values, e.g. the determinand "Discharge" can only be observed in medium "Water".

In fact, OWL only supports limited relationship expressions between instances, e.g. *owl:differentFrom* and *owl:sameAs*. User-defined instance relationships are not allowed

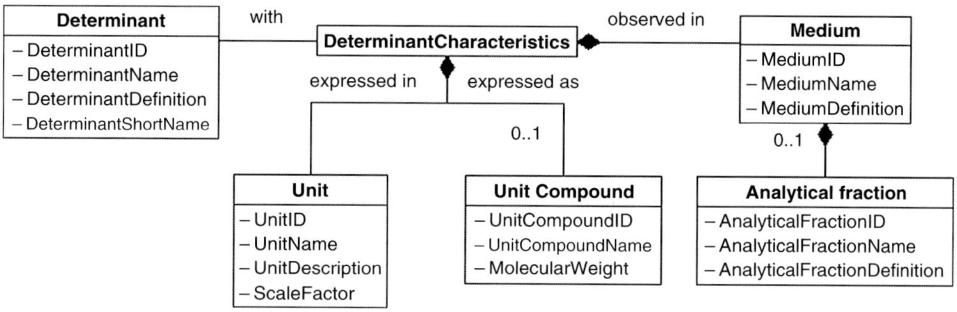

Fig. 9.2. EGV representation of determinands and associated classes.

in the OWL syntax, which makes the expression of instance relations difficult to model in practice.

In any information model design, there are several data modelling choices. For example, whether to represent data instance relationships as data class, or type relationship. There may also be redundancy in the data representation, e.g. data model redundancies may exist in the relational database, and SQL supports different ways to express a *join* between relational data tables. SQL can use normalisation to prevent data redundancy. However, it may not be clear for domain developers how to capture knowledge in the domain, and how to represent it. Other modelling choices include how to represent class relationships in a domain; whether to represent concepts into set or *has-a* relationships; or to represent the same concepts using class inheritance or *is-a* relationships.

Hence, we could, for example, also represent a fragment of the domain as follows: "Nitrate" and "Nitrite" are chemical determinands observed in the inland water domain. They have a common set of properties such as: "determinand name", "determinand ID", "definition", and "chemical formula". The determinands may be grouped into subset according to their inherent chemical categories and properties. We can have a determinand subset called "Nitrogens_Oxided", which contains the determinand values "Nitrate" and "Nitrite". Another subset "Nutrients" has also the value "Nitrite", as the subsets in practice may overlap.

The statement above can be expressed in an Ontology with two distinct conceptualisations: *inheritance* or *subset*.

If we define that class "Determinand" stands for the set of all observed determinands in the inland water domain, and class "DeterminandList" stands for the set of determinand groups, then class "Nitrogens_Oxided" is a subclass of "DeterminandList".

In the inheritance case, "Nitrate" and "Nitrite" can be abstracted as the disjoined subclasses of "Nitrogens_Oxided". Semantically the inheritance hierarchy implies that a class can inherit all properties from its super-class, i.e. "nitrate" is a "determinandList", although it leads to confusion because "nitrite" and "nitrate" are instances of determinand. The redundancy definition benefits from further definition of the determinand collection at an even lower granularity level, e.g. "nitrite" can be defined as a collection of varied compounds. An example OWL representation is shown in Figure 9.3.

In Table 9.2 it is shown that the class "Determinand" has an instance of "nitrate" with a set of properties (formula, definition, etc.). "Nitrate" is a subclass of "Nitrogens_Oxided" and is defined by the property "hasIdeterminand" having exactly the value of the instance "nitrate".

In the alternative understanding of the subset case, the "nitrite" and "nitrate" can be simply defined as the instance of "Determinand", while "Nitrogens_Oxided" is defined exactly as an enumeration value class consisting of "Nitrite" and "Nitrate", which is shown in Table 9.3.

Both the models shown in Tables 9.2 and 9.3 are correct in the sense of having a correct OWL syntax. However, they represent variations in the interpretation of the

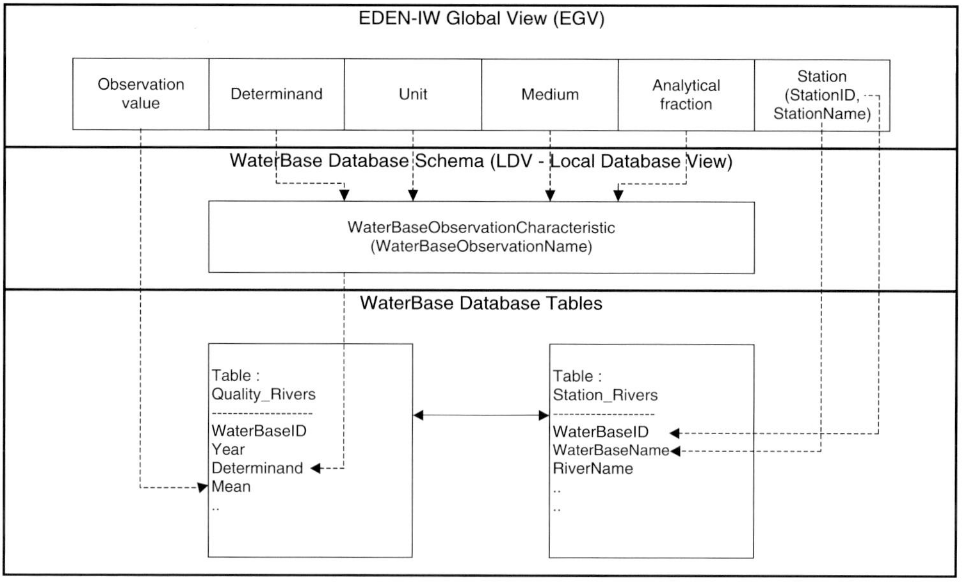

Fig. 9.3. Semantic mapping example for the waterbase database.

Table 9.2. Determinand list modeling in inheritance relation

```
    <owl:Class rdf:ID="Nitrate">
      <rdfs:label xml:lang="en">nitrate</rdfs:label>
      <owl:disjointWith rdf:resource="#Nitrite"/>
      <rdfs:subClassOf>
        <owl:Class rdf:about="#Nitrogens_oxided"/>
      </rdfs:subClassOf>
  <owl:equivalentClass>
    <owl:Restriction>
      <owl:hasValue>
        <Determinand rdf:ID="nitrate">
            <DeterminandID>19</DeterminandID>
            <DeterminandDef>Nitrogen in the form of NO3- </DeterminandDef>
          <EDENTermID>1014</EDENTermID>
          <ChemicalFormula>NO3-</ChemicalFormula>
          <DeterminandName>nitrate</DeterminandName>
        </Determinand>
      </owl:hasValue>
      <owl:onProperty rdf:resource="#hasIdeterminand"/>
    </owl:Restriction>
  </owl:equivalentClass>
</owl:Class>
```

Table 9.3. Determinand list modeling using the subset relation

```
<owl:Class rdf:ID="Nitrogens_oxided">
  <rdfs:subClassOf rdf:resource="#DeterminandList"/>
  <owl:equivalentClass>
    <owl:Class>
      <owl:oneOf rdf:parseType="Collection">
        <Determinand rdf:ID="nitrate">
          <ChemicalFormula rdf:datatype="http://www.w3.org/2001/XMLSchema#string"
          >NO3</ChemicalFormula>
          <DeterminandName xml:lang="en">nitrate</DeterminandName>
          <owl:differentFrom>
            <Determinand rdf:ID="nitrite">
              <DeterminandName xml:lang="en">nitrite</DeterminandName>
              <ChemicalFormula xml:lang="en">NO2</ChemicalFormula>
              <DeterminandID rdf:datatype="http://www.w3.org/2001/ XMLSchema#
                string">18</DeterminandID>
              <owl:differentFrom rdf:resource="#nitrate"/>
            </Determinand>
          </owl:differentFrom>
          <DeterminandID rdf:datatype="http://www.w3.org/2001/XMLSchema#string"
          >19</DeterminandID>
        </Determinand>
        <Determinand rdf:about="#nitrite"/>
      </owl:oneOf>
    </owl:Class>
  </owl:equivalentClass>
</owl:Class>
```

domain knowledge from different viewpoints. The modelling of domain Ontology is not a straightforward process leading to a single monopoly result. The representation of a domain Ontology model may vary depending on several factors including the expressivity of the Ontology language, the scope of the domain, requirements, the application commitments, and the Ontology development process.

9.4.5 Supporting Heterogeneous Instances: Unit Ontology

In the real world, different representations or structures for instances are also possible, e.g. concentrations of determinands can be measured and expressed in units from different measurement systems. Hence, domain models need to consider the following issues: (a) how they capture heterogeneities in real-world entities (b) how these are related to each other (c) whether or not they capture and represent relationships between these heterogeneous instances in the domain model. Generally, many semantic models are not expressive enough to support general data transformation rules and rule-based processing of the semantic data.

Additional design requirements relate to the ability to convert between different units. One problem with units in the relation conversion is that it is cumbersome to define conversion factors for all the possible combinations. The solution is to define a set of basic unit classes (weight, length, time, etc.) with instances in the EGV model. For each instance the scaling factors (offset and scale) have been defined relative to the SI unit. More complex core units are defined using the basic unit classes. A "FluidConcentration" unit is a subclass of "ConcentrationUnits" and is defined by having a numerator from the "WeightUnits" and a divisor from the "VolumeUnits". Different unit instances may now be compared according to the class type they are instances of. "ConcentrationUnits" is a subclass of "FractionUnits", which is specified to have both a numerator and a divisor.

Comparison of different instances of "ConcentrationUnits" may now be applied by using a general rule for all "FractionUnits", and using the scaling factors for both numerator and divisor.

9.4.6 Development and Maintenance of the EDEN Global View Ontology

The EDEN-IW Ontology development was started when the project began. There are well-defined processes and methodologies for Ontology development (OntoWeb-D1.4, 2005; Noy and Klein, 2004). Processes differ depending on whether or not the Ontology is developed from scratch, the ontologies are cooperatively constructed, or the ontologies are re-engineered from existing ontologies (OntoWeb-D1.4, 2005).

EDEN-IW focused on building an Ontology from scratch. Whereas an Ontology could be constructed using low-level tools as text-editors and Ontology validators, Ontology development environments ease the development. They include visual tools for graphically creating and editing ontologies, and exporting representations for online use by application processes. Because of the requirement of EDEN-IW to focus on XML type ontologies, EDEN-IW chose DAML+OIL as the Ontology language. Tools that support DAML+OIL developments include OILEd (http://oiled.man.ac.uk) and Protégé (http://protege.stanford.edu). Of these, Protégé was considered to be the most mature. However, the newer versions of Protégé do not any longer support DAML+OIL, but support OWL instead.

Ontology creation

In developing Ontologies with Protégé, Noy outlines the following steps for engineering ontologies (Noy and Klein, 2004):

- *Determine the scope of the Ontology*: The scope of the Ontology is for inland water including lakes and rivers. Seas and oceanic water measurements were considered to be outside the scope, although the scope of the Ontology could be expanded at a future stage to include these.

- *Consider reuse*: It can be more effective to reuse an existing domain Ontology rather than to construct one from scratch. At the start of the project, an XML-based Ontology for the IW domain was not available, so the domain experts in the project created one.
- *Enumerate important terms*: Define the concepts we need to talk about, their properties and what do we want to say about the terms, e.g. water medium and measured chemical parameter.
- *Define classes and the class hierarchy*: Associate concepts with classes (collection of concepts with similar properties), e.g. Determinand, Medium, and Unit. Define a taxonomic hierarchy to relate classes of related sub-types and super-types, e.g. the super-classes are EDENGlobalConcept and Determinand, Medium and Unit are sub-types of these.
- *Define properties of classes*: Describe attributes of instances of the class and the relations to other instances, e.g. the Medium concept class has attributes of Name, ID, and Definition. Simple properties (attributes) contain primitive values, e.g. ID (strings, numbers), but properties that are more complex may link to other classes.
- *Define constraints*: Property constraints (facets) describe or limit the set of possible values for properties, e.g. an ID property is defined as a unique Integer Identifier.
- *Create instances*: For example, Aluminium is an instance of the Determinand class.

The whole development process of the Ontology may be described as an iterative process of refinement. This is done by the exchange of domain knowledge between the domain experts (for inland water) and the Ontology (agent infrastructure) developers. EDEN-IW used a combination of bottom-up and top-down methodology to define the more salient concepts first derived from the instance data, and then to generalise and specialise these in the domain model. Note that this creation process does not consider the use of the metadata to operate on other different data representations, e.g. to obtain more relevant search data instances, which are not in the metadata representation. If this is the case, then additional steps are needed to map or relate the metadata instances to specific (database) source instances.

An important aspect in defining the classes and the class hierarchy is to be precise about the actual relationship between the different concepts. The following example illustrates this. When analysing environmental observations from a river station, a number of characteristics that are related to the station will be of interest. These may comprise the size of the catchment contributing to the flow of water passing the station and the population living in the catchment. In practical implementations of inland water databases, these characteristics may often be gathered and stored with the reference to the station ID. An implementation of the "CatchmentArea" and "Population" as properties of the Station class are conceptually not correct, and would result in a revision of this particular class hierarchy.

The more appropriate approach would be to link the station to a position on a river stretch. Such a point will have an associated catchment. The catchment is a surface and will have its area as a natural property. The population (or population density) is then an observation linked to a spatial object, which represents a surface or a volume.

There are two key challenges in using ontologies once they are created:

1. How to maintain the Ontology
2. How to orientate the Ontology to different sets of applications, and different sets of types of users

Each of these challenges is discussed in turn.

Ontology evolution

It may be supposed that a domain Ontology model should be created and iteratively edited until it is complete, expressive, and accurate. Only then is it fixed as a Knowledge Interface to subsequently use for all users and applications. This is seldom the case in practice – ontologies are likely to evolve. Ontology evolution can be defined as the timely adaptation of the Ontology as well as the consistent propagation of changes. A modification in one part of the Ontology may generate subtle inconsistencies in other parts of the same Ontology, in the instances, dependent ontologies and applications. This variety of causes and consequences of the Ontology changes is discussed in Klein et al. (2002). Ontologies are living and have a maintenance phase where parts may change. The main sources of change are (OntoWeb-D1.4, 2005):

- *Structure-driven change discovery*: Exploits a set of heuristics to improve an Ontology based on the analysis of the structure of the Ontology. For example, if all sub-concepts have the same property, the property may be moved to the parent concept.
- *Data-driven change discovery*: Detects the changes that are induced through the analysis of existing instances. For example, if no instance of a concept C uses any of the properties defined for C, but only properties inherited from the parent concept, we can assume that C is not necessary.
- *Usage-driven change discovery*: Takes into account the usage of the Ontology in the knowledge management system. It is based on the analysis of the users' behaviour in two phases of a knowledge management cycle: in providing knowledge by analysing the quality of annotations, and in searching for knowledge by analysing the users' queries and the responses from the knowledge repository. For example, by tracking when the concept was last retrieved by a query, it may be possible to discover that some concepts need to be modified.

In order to manage Ontology evolution in practice, we adopt an approach where each Ontology version is assigned a version identifier that includes a date part and make them available via a Web server. As new versions of ontologies evolve, they are added to the Web-server. Applications that use these Ontologies must take care to specify the version of the Ontology they load at run-time in order for the application to be synchronised with the Ontology definition. It is also possible to organise the Ontology evolution in terms of a core model with different extension ontologies that are interlinked using XML name-spaces. The extension is linked to a previous version of the core Ontology, with defined difference concepts.

Ontological commitment
It is difficult to construct a single Domain Ontology that would be suitable for all domain applications and for all users in that domain, because these can vary and change at different rates leading to a very complex and frequently changing Ontology. For this reason, it is proposed that ontologies are partitioned into layers. There are several different layered models of ontologies, e.g. the core conceptual Ontology layer, and the application specific layer, where applications explicitly specify their particular commitment to the Ontology. In the EDEN-IW project there is a core Ontology model called the EDEN-IW Global View or EGV, and a set of Local Database Views or LDV. The latter layer is where the specific database application instances make a commitment to the global Ontology model. In the LDV it is also defined how the instances link (or extend) the core concepts to the local concepts. Another example of a different type of commitment would be a foreign language specific commitment. Here terms in the core Ontology in e.g. English are mapped to another language such as French or Danish.

Two experimental ontologies have been developed in the later part of the project. The first Ontology is a Unit Ontology oriented toward the translation between units; the development of the second Ontology is oriented towards a common knowledge base of relations between determinands.

9.4.7 Ontology Provenance

An important challenge for the Semantic Web is that whilst a consensus regarding the concepts, structure and scope of a model can be achieved within a community, many different communities can promote their local Ontology model to a global community as being "the" domain model for a particular domain. This increases the risk of a lack of interoperability (the opposite of one of the main reasons for using ontologies) between different ontologies within the same domain, and the risk that a poorly formed or defined Ontology for that domain could take hold.

One way out of this conundrum is to prepare or reference a domain Ontology to terms that have International provenance, as far as possible. However, such internal glossaries of domain terms are often not in a machine-readable form. Furthermore, a consensus amongst experienced domain experts together with all the relevant stakeholders, who commit to use the Ontology, and a process of refinement and peer-review is needed to create and maintain the Ontology.

Within EDEN-IW, a lightweight IW domain Ontology in the form of multiple international standard thesauri or glossaries of accepted terms such as GEMET (GEneral Multilingual Environmental Thesaurus), T-REKS (Thesaurus-based Reference Environmental Knowledge System), and EARTh was used to provide provenance (see Chapter 10). To provide a machine-readable access to these terms, glossaries were interlinked and encoded in XML. Whilst the definitions or descriptions of the terms are in English and are not machine-readable, the term names along with synonyms are machine-readable. Each concept in the EDEN-IW common or global IW Ontology was linked to one or more terms in the glossaries via identifiers.

9.4.8 Ontology Services

Ontology services are needed to allow applications and services to load and parse the EDEN-IW Ontology models, in order to support querying and retrieving of the local database data. The Ontology services are implemented as Java applications that were developed using Jena, which is a Java framework for building Semantic Web applications developed by HP (available from http://jena.sourceforge.net/). The current version of Jena in 2005 provides a programme environment for RDF, RDFS, and OWL, including a rule-based inference engine. At the start of the project, the focus was on DAML, supported in an older version of Jena, as this was the most mature semantic model. As the project progressed, support for OWL became more mature, and as a result the project focused more on OWL.

Ontology access

The objective of building the Ontology services is to improve the reusability and efficiency of the Ontology access from different users. Although the initial focus was on content harmonisation, further developments have focused on use of the Ontology mode via multilateral user and multiple application viewpoints. Using the EGV concepts and relations, EDEN-IW users can easily build the user query according to the specific interests of an information fraction in the knowledge domain. EGV instances were created in order to maintain the value mapping within the particular EGV class, e.g. mapping between "determinandID" and "determinandName". Additionally, aggregation relations were introduced into the EGV model to give the explicit definition of the particular domain logic. For example, "Observation" is an aggregation of several other classes such as "Determinand", "Medium", and "AnalyticalFraction".

Ontology mapping: from local database view to global Ontology view

The LDV model consists of database schema and local logic concepts, derived from the EGV concepts, so that the LDV can be mapped to EGV. The representation of the LDV model in DAML adopts several generic rules to ensure the scalability of the LDV model and reusability of the Ontology service. Each table in a database is abstracted into a subclass of the common super-class concept of Table. Each column is abstracted into a sub-property of the common super-property field. A Foreign Key relation is presented using the DAML tag of "SamePropertyAs". The Primary Key is marked using the DAML tag of "UnambiguousProperty". By browsing the DAML tags, the Ontology services can inference about the data model.

Mapping relationships need to be set up between the EGV and LDV data views. "SamePropertyAs" was used to mark the semantic equivalent properties crossing the two models. The mapping includes three categories:

1. *Direct mapping from database column*: The direct mapping is applied under the condition that the EGV property has a direct equivalent column in the database schema. Thus, no processing is needed to precede the logic or mathematic conversion.
2. *Value conversion*: Value conversion is applied when the EGV property has the same semantic meaning as LDV property, but the direct mapping could not be established

due to the problem of different coding format and value representation between EGV and LDV terms. In this case, a local logic concept was introduced to inherit the EGV concept and to provide the value mapping or processing. Direct mapping is needed to map logic property to the database column. For example, due to the different ID coding of EGV determinand and IOW determinand, "IOWDeterminand" is created in IOW LDV as the subclass of "Determinand" in EGV, "IOWDetermint" consists of "DetermintID", which comes from EGV "value", and "IOWDeterminandID".

3. *Logic conversion*: Value conversion does not resolve all mapping problems. Logic conversions are required in the case of more complex logic mapping between LDV and EGV concepts. The complex LDV concept has different semantic meanings that can be mapped to an EGV concept directly. Normally, it can be represented as the logic combination consisting of several EGV concepts. For example, in NERI LDV, a local logic concept "NERIObservationCharacteristic" was used to abstract the concept of "NERI Inland Water Observation" that contains the value combination from several EGV concepts such as "Determinand", "Medium", "AnalyticalFraction", and "Unit".

From the Cardinality view, we can mark direct value mapping as a one-to-one (1-1) mapping between two Ontology views, whereas logic conversion mapping involves more complex relations of one-to-many mappings. An example ontological mapping for the WaterBase database is illustrated in Figure 9.3.

An automatic generation of the local view Ontology based on, e.g., a database system file would be a beneficial target. However, such a goal is not easily achievable. The process of performing the first mapping of a database to the EGV will always have to include people with knowledge about the database structure and the concepts behind it. A simple element such as a field label does not necessarily contain a term from a natural language, and even if it did, the interpretation of the concept would still have to be verified.

In the EDEN-IW project, the Local-centric approach (Calvanese et al., 2001) is adopted. Here concepts of the local Ontology are mapped to a corresponding view in the global Ontology. The development of semantic mapping is conducted as in the process described in Figure 9.4.

The mapping process for relating local to global terms to be able to use queries expressed in global terms to access local terms is defined as follows:

- A part of the local schema for export is selected for the semantic mapping.
- This schema is expressed in an Ontology format (OWL/DAML+OIL).
- Any concepts and properties that have equivalent meaning across both LDV and EGV are identified.
- A semantic analysis is conducted to determine the mapping relationships.
- One-to-one semantic mappings are marked as a direct mapping.
- One-to-many semantic mappings are marked as logic conversion mappings where a LDV concept is viewed as a collection of EGV concepts linked to a particular EGV relationship. An intermediate class is defined in LDV.

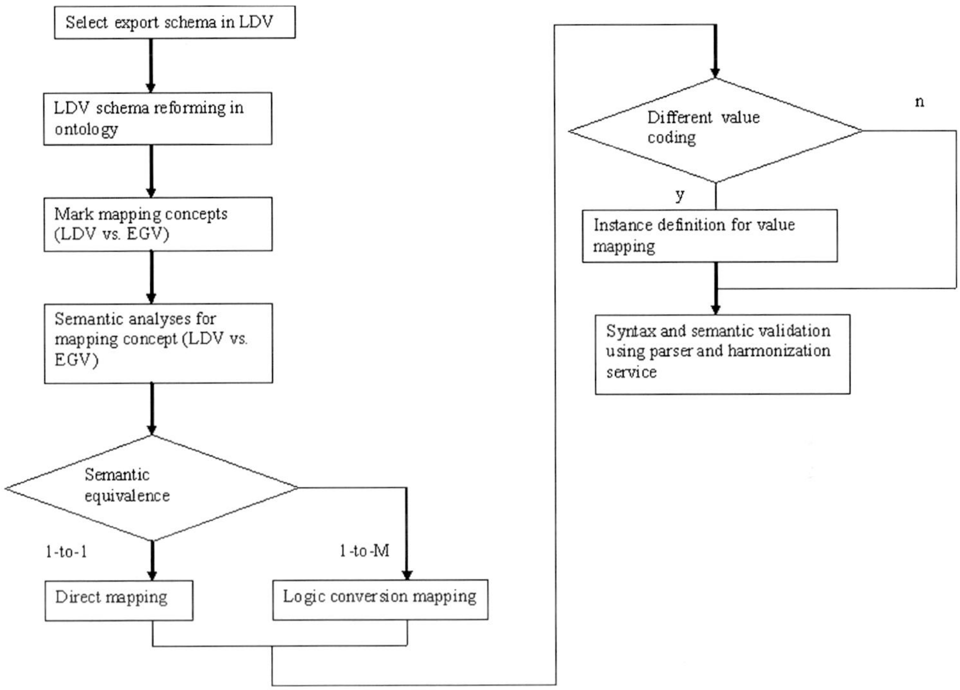

Fig. 9.4. Mapping process for relating local to global Ontology terms.

- The value-coding format cross EGV and LDV for those mapping concepts is compared. If their formats are different, a corresponding instance value mapping is defined in the Ontology.
- The syntax and semantic validation is implemented in a message handler and parser.

The semantic mapping across the global and local Ontology views supports the query transformation between EGV and LDV by giving explicit descriptions of terms, views, and instance value translation. SQL query can be generated based upon the LDV terms and executed on one or more local databases. The SQL query results are returned and mapped back into EGV terms in order to be presented to the user. To support this, the LDV includes information to link and reference the EGV terms to the corresponding database table structures such as table names, columns, data-types, primary, and foreign key relations.

SQL generation

The application builds the SQL statement according to the user input query that is converted to RDF. An example is shown in Table 9.4.

The query statement gives an SQL-like query structure that consists of two sub-sets for the query arguments and the constraints statement. The former is represented as a RDF

Table 9.4. Example of query in RDF

```
...
<fipa:argument>
<EDEN:DeterminandID><rxsd:integer><rdf:value>4</rdf:value></rxsd:integer>
  </EDEN:DeterminandID>
<EDEN:MediumID><rxsd:integer><rdf:value>*</rdf:value></rxsd:integer></EDEN:
  MediumID>
</fipa:argument>
<fipa:result>
...
<EDENSchema:selection-result ID="0">
```

tag argument and the latter one as a constraint. This XML/RDF representation hard-codes the semantic logic of a user query into a specific structure. Each user query asks for the value of one or more properties or columns defining the constraints. The mapping SQL statement is of the form:

SELECT DISTINCT param.code_param,stations.localisation
FROM parametres, stations, mesures
WHERE ((param.CODE_PARAM=mesures.CODE_PARAM)
AND (mesures.CODE_STATION=stations.CODE_STATION)
AND (param.CODE_PARAM=1311))

The SQL statement can be related to the XML query structure as follows:

1. The result set of XML/RDF is semantically equivalent to the SELECT in SQL.
2. The argument set of XML/RDF is semantically equivalent to the WHERE clause of SQL.
3. Two inputs are needed to define the SQL statement: The table names where the columns belong, and the relations to join those tables.

In points 1 and 2, the EGV terms and values in the RDF query can be translated to the LDV terms and values directly, using the Ontology service. By browsing the LDV model, the SQL building service can find the related table name for the particular columns. Then, the only question is how to join these tables together and form the WHERE section in SQL. A graph algorithm helps to calculate a join between tables. Each table can be seen as an individual node in a graph, and each foreign key is the arc to link different nodes together. The calculation of the join then becomes the determination of a path connecting given nodes.

9.4.9 Content Harmonisation

Often data from multiple data sources cannot easily be compared, because the data represents different values. This could be due to whether or not the measurement system has been calibrated recently, or because the data has been averaged differently. Additional metadata to record the provenance of the data from the measurement source is needed, including metadata about the measurement technique and any post-processing of the measurement data. This is needed in order to do a true comparison, e.g. a unit conversion may be needed to equate measurement data in different units.

The Ontology service can capture disparate syntactic representations and semantic interpretations of domain knowledge in different lateral ontologies. The different lateral ontologies can be translated because of the semantic mapping of each lateral Ontology into a common EGV. Currently, data harmonization is handled in the Task Agent rather than captured in the declarative semantic metadata model itself.

9.5 DISCUSSION

International data sharing programs will necessarily be based on agreements of delivery. Due to cost restrictions and differences in monitoring traditions, such agreements will tend to focus on "lowest common denominator" approaches. The EDEN-IW approach gives instead the opportunity to exploit the sum of knowledge and data. The system is able to do this by using a semantic model that is open and scalable in several directions:

- New determinands and concepts can be added to the EGV Ontology and are then available to other parts of the system: The simple addition of a new determinand or database does not require any changes in user interfaces or other parts of the agent system.
- New application areas can rely on existing Ontology elements and thus focus their efforts on developing their domain specific concepts and applications.
- New user interfaces and processing agents can be developed by anybody. The only requirement is that the interfaces conform to the external EDEN-IW system interfaces, e.g. the EGV model.

In addition, the design of the EDEN-IW Ontology model through its partitioning can isolate the development of the software of the information processing system from the domain specific knowledge.

The EGV model and the associated application commitment models are represented using the W3C RDF, DAML+OIL, and OWL-based models. A set of intelligent software agents has been created within the EDEN-IW project to allow users to make a rich set of data and metadata queries, and to support semantic data exchange and reasoning between heterogeneous database resources, metadata repositories, and user applications (see Chapter 8). These agents are implemented using the FIPA agent forum standards implemented in a Java agent tool-kit called JADE and a Java based framework called Jena that allows agents to parse RDF, DAML+OIL and OWL messages. Queries are expressed in the EGV Ontology. These are mapped into the LDV models of the corresponding database resources and then to SQL queries. The same agent services will also translate the SQL responses from the database back into the common language (EGV) terms.The system has demonstrated the integration of four heterogeneous IW databases.

Chapter 10

QUADRILINGUAL LIST AND TERMINOLOGY REFERENCE SYSTEM

B. Felluga and P. Plini

10.1 INTRODUCTION

In a multinational collaboration project, such as EDEN-IW, the terminology was necessarily foreseen to be multilingual. In principle, no preference had to be given to any of the involved languages: Danish, English, French, Italian, Portuguese, or any other language. The representation of the language involved depended, in principle, on the language of the databases willing to participate to the project by providing terminology and inland water data. Incidentally, during the course of the project, Italy contributed with most of the terminology work, but did not provide water data; Portugal dropped the participation at an early stage of the project, and other countries/languages were not involved because of time and effort costs.

During the project, English was used as the common working language. English was thus the "*de facto*" preferred language. For the formal terminological representation of terms, British English was chosen as default and display language. Thus, English was the primary source language of English-based terminology and became the secondary source language for the non-English-based terminology.

The alignment of British English terminology and the North American English terminology was not foreseen in the project, in spite of the usefulness of the alignment of some important set of terms, including the measurement units. This issue is a matter of politics, policy, and cost-benefit analysis.

During the first steps of EDEN-IW project, most of the attention was on the design of the ontology model, the choice of the information technology platform, and the identification and involvement of the data providers in different countries (and languages). For this reason, the basic terminology needs of the EDEN-IW project were defined in detail somewhat later than the start of the project, as operationally, it could not be different. Meanwhile, it was felt that the needs of terminology had not to be restricted to the language of the queries in the EDEN-IW system, which was foreseen to be in the order of a few hundred terms. It required precision of the IW terminology as a general reference-specific language for the lay user, in the order of no less than one thousand terms.

In this context, it should be noted that an aim of the EDEN-IW project was to ease the sharing of knowledge and water quality data between heterogeneous databases. The heterogeneity was in several areas: in semantics, in the ontologies used, in the information technology applied, and also in the use of linguistics and terminology. Furthermore, in the project, a sub-task of the CNR task was the comparison of the EDEN-IW terminology with the environmental multilingual thesaurus of EEA. At European level, the purpose of this task was to move towards availability of harmonised multilingual environmental terminologies.

Consiglio Nazionale delle Ricerche (Italian National Research Council) (CNR) was familiar with compilation and upkeeping of multilingual environmental terminologies. Through a multi-annual effort it had succeeded in collapsing the Multilingual Descriptor System (MDS) of the European Commission (1983) and several monolingual thesauri and lexica in a unique thesaurus named GEMET (EEA, 1999). This multilingual environmental thesaurus, consisting of 5300 terms in 12 languages, was compiled for EEA by CNR together with UBA, UNEP, US EPA, and some additional national institutions. It was also adopted as the core of the US EPA Terminology Reference System (2000). Thus, a multilingual unique reference terminology for the environment was already available.

The rationale of this work was that the basic concepts for the description of environmental issues do represent specialised knowledge. This specialised knowledge can be identified by a unique (non-intelligent, neutral) identifier, which is shared in the common communication level among scientists and operators of the environmental information domain. In practical terms, this happens, for instance, in scientific congresses through the means of a so-called "relay language".

Obviously, for the same concepts, different definitions could be collected from different glossaries, thesauri, handbooks, and so on. These definitions would be in different languages, in different data dictionaries, and in different contexts, e.g. juridical vs. ecological.

It was also clear that one concept could have different terminological (linguistic) representations in the same language. It could have different terminological (linguistic) representations in different languages. It could also have different terminological (linguistic) representations in databases for different purposes, e.g. library vs. factual. Even in the same type of databases, the knowledge organisation could also lead to different terminological (linguistic) representations.

Nevertheless, this did not pose problems with the "consensual" utilisation of terms in the daily information exchanges between scientists and environmental operators. It is a potential problem when

- high precision is required, e.g. in various contexts of juridical language.
- a misconceptual linguistic representation would lead to ambiguities, homonymies, etc.
- a term is mistranslated, leading to an incorrect equivalence between concepts in different languages.

All these not only illustrate the complexity and difficulty of the terminology work, but provide also a rather clear, defined, straightforward operational frame within which to cope with the multilingual semantic and terminology issues.

Summarising, the CNR was aware of the need to compile a terminology reference document that could be used by different operators in different countries, not too precise to become a burden with too subtle details, but sufficiently precise to allow consensual practical utilisation of a sound scientific language. The expected benefits were the use of a unique terminology system leading to economical use of resources and efforts.

At this point, it seems worth to take into account that the multilingual terminology work, in particular the compilation and maintenance of a multilingual thesaurus of several thousand terms, is a tremendously energy-time-money- and resource-consuming work. Without suitable, continuous investments, a thesaurus can be neither maintained nor updated, and in a few years, it becomes obsolete. On the one hand, the investments are often bound to contracts; the contracts depend on the agencies' multi-annual investment plans, which depend on the agency policy, governmental politics, and, at large, strategic trends. GEMET was a contract-based product and not an institution-endorsed one, which, during the years following its publication, was translated in other languages through voluntary national contributions but neither quality control was applied nor was GEMET extended. On the other hand and independently, since 1999 CNR pursued its work on the thesaurus, in a first step by editing a new thesaurus, Environmental Applications Reference Thesaurus (EARTh) (2000–2002) and later by developing this thesaurus to a much larger extent (10 000 terms) with the addition of terms from the IUGS multilingual thesaurus, the EDEN-IW project, and other relevant sources (2004–2005). In 2004, EARTh became the core of the Terminology Reference System (TRS) of the Italian Ministry of the Environment.

Thus, by 2001, an inland water multilingual terminology was already available in EARTh. However, it was scattered in a broad conceptual context, not exactly oriented in the direction of the EDEN-IW project. On the other side, relevant multilingual thesauri and glossaries of the water domain existed and were available either through institutional contacts or the Web. IOW had already published and was maintaining the "Thesaurus de l'eau", and the UNESCO *International Glossary of Hydrology* had been published in 1999. Other documents contained further terminology to be taken into account for the EDEN-IW project, like the Water Framework Directive, Waterbase, and so on.

10.2 THE QUADRILINGUAL LIST OF DETERMINANDS

10.2.1 Introduction

This section describes the "EDEN-Inland Waters Quadrilingual List of Determinands" (shortly, "the List"), in English, Danish, French, and Italian. The List is the Basic Quadrilingual Query Language of EDEN-IW. It presents multilingual alphabetical sequences of the Determinands as they will appear to the EDEN-IW user on the video screen: the user will choose the searching language (one of the four languages) and thus have available the complete alphabetical list of the determinands in the chosen language.

The List should be considered from the semantic and terminological point of view. The basic reference and working language was British English, i.e. the UK English, while the source languages for the List were Danish and French. A Danish and a French institution have provided and agreed upon a common proposal for the first determinand list of about 300 concepts/terms. Moreover, at the onset of the Project, a limited number of terms provided by a Scottish institution have been taken into account. This list named "Cross-Mapping Table" has provided the basic material for the analysis and compilation of the present List of Determinands. Terms from the Priority Substances of the Water Framework Directive have been added to this List in order to adjust the semantic coverage of the List to the needs of users of the European Union.

There is an historical and operational difference in the richness of water terminology in Denmark and France, owing to the size of the countries, the different implementation approaches, and other aspects as well. The List of Determinands reflects, to a certain extent, this difference.

The List provides an example of a Multilingual Terminology Reference System, which is somewhat less developed than a thesaurus, though significantly more developed than an Authority List. The methodological solutions adopted to compile the List (e.g. the permutation of terms) are rather original. However, they need further improvements both at the semantic and terminological level. This is particularly important for the difference between terms that are groups of substances, like "metals" versus the individual species of substances belonging to them, like the individual "metals". At the theoretical level, a solution to this problem exists, largely implemented in practice in the thesauri by means of the hierarchical handling of concepts/terms as Broader and Narrower Terms. Nevertheless, in the List two aspects prevent an easy solution to this issue:

1. A certain fuzziness is still persisting in the reference chemical nomenclatures such as the CAS Registry Number, the EINECS Numbers, etc.
2. The different approaches implemented in different water institutions of different countries for the handling of determinands, from many different viewpoints: regulatory, strategic, technical, informational, terminological, linguistic, and so on.

Thus, needless to say, the List can be subject to amendments and extensions, following the feedback from involved partners from different countries and institutions.

10.2.2 Parameter, Determinant or Determinand?

In order to avoid ambiguity and improper use, the definitions of "determinant" and "determinand", two terms of broad usage that should not be confused, were checked in authoritative dictionaries. According to *Merriam Webster*, a determinant is "an element that identifies or determines the nature of something or that fixes or conditions an outcome" (other dictionaries like *Oxford, Encarta*, etc. will provide definitions with the same meaning). This meaning is clearly active and transitive. Conversely, for the purpose of EDEN-IW project, a chemical that is determined with regard to occurrence in a water body as indicative of pollution is a determinand, i.e. something that is, or can be, or will be determined by an analytical method or procedure.

In this context, a determinant determines something else, and a chemical, by its occurrence, can determine the pollution of a water body, thus being, in a different semantic context, a determinant: in such a case, it would be incorrect to call it determinand.

In the Water Framework Directive, the relevant inland water bodies where determinands are determined (identified, analysed, measured, monitored, etc.) are:

1. groundwater
2. lakes: lakes and reservoirs
3. rivers: rivers and canals.

Incidentally, the data collected for Eurowaternet (and thus potentially available in the publicly available Waterbase) are restricted to a limited set of determinands:
R = determinand requested for rivers stations.
L = determinand requested for lakes stations.
G = determinand requested for groundwater stations.

In the EDEN-IW reports, the term "determinand" was frequently used as synonym of "parameter" and vice versa. In the present text, the term "determinand" is preferred, although it is possible that the ambiguity between "parameter" and "determinand" persists also in some texts of the present publication.

10.2.3 Why a Determinand List?

The EDEN-IW user supposedly needs a series of elements to combine in order to formulate his/her query. To this purpose, NERI and IOW have selected the lists of elements shown in Table 10.1, where the prevalence of determinands of chemical and biochemical nature can be seen.

Table 10.1. Different elements to be combined in the user's query and their numbers

Element	Type	No. of Terms	Example
Determinands	Bacteriological	4	temperature, nitrate, etc.
	Biochemical	10	
	Chemical	246	
	Physical	8	
Analytical fraction		9	inorganic, organic, etc.
Units		73	ng/l, hour, etc.
Media		7	water, fish, etc.
Value attributes		5	greater than, less than, etc.
Subjects		13	ground water, river, etc.
Methods		8	5 days, in situ, etc.
	Total	383	

The present List consists of the list of determinands, complemented with the lists of Analytical Fractions, Units, Media, Value Attributes, Subjects, and Methods. It presents a complete equivalence between the languages, including Italian.

10.2.4 Development of the List

Materials: terminology sources
The sources of determinands and their synonyms are shown in Table 10.2.

As regards the synonyms of the chemicals, the source or the source of sources used for checking the proper name, the synonyms, the formulas of pesticides and chemicals, are shown in Table 10.3:

These sources have been utilised to check the orthography of the term representations in the various languages.

Methods and procedure
The steps performed during the implementation of the List included the usual terminological handling of concepts/terms and their synonyms, variants, and accessory specifications. The latter are particularly important identifiers for chemical substances. Although the working language has been UK English, which provides a relevant series of synonyms of international use, the terminological experience of the French IOW contributed consistently to the enrichment of the List. Particular attention was paid to clarify the terms that could potentially pose problems to the user. Internet sources were used extensively to ensure a good control of semantic coherence between terms, their specifications, and their definitions, as well as for validating the choice between Entry Terms and their Synonyms. Eventually, a Data Dictionary of the EDEN-IW Quadrilingual List of Determinands database was implemented. Normalisation, translation, and e-editing completed the work.

10.2.5 Results

The List in the context of EDEN-IW Terminology Reference System
The present List is part of the overall EDEN-IW TRS. It is useful to remember that the whole TRS consists of

Table 10.2. Sources of determinands

Sources of determinands	URL
NERI Determinand List	http://www.dmu.dk/International/About+NERI/
IOW Determinand List	http://www.rnde.tm.fr/anglais/pr/prpa0014.htm
SEPA Determinand List	http://www.sepa.org.uk/index.html
WFD Priority Substances List, version 2004	http://europa.eu.int/comm/environment/water/water-dangersub/pri_substances.htm

Table 10.3. Source or Source of sources of terms, synonyms and formulas of pesticides and chemicals

Source or Source of sources for terms, synonyms and formulas	URL
Agency for Toxic Substances and Disease Registry	http://www.atsdr.cdc.gov/97list.html#bookmark01
ChemFinder.com	http://chemfinder.cambridgesoft.com/
ChemIndustry.com	http://www.chemindustry.com/index.asp
Compendium of Pesticide Common Names	http://www.alanwood.net/pesticides/index.html
Dr. Ehrenstorfer GmbH	http://www.analytical-standards.com/
Environmental chemistry, safety, MSDS	http://www.library.adelaide.edu.au/guide/sci/Chemistry/env.html
EPA Pesticide Regulation Notice 97-5 (September 1997). Use of Common Names for Active Ingredients on Pesticide Labeling.	http://www.pestlaw.com/x/notice/epa/PR97-5.html
IPCS INCHEM – Chemical Safety Information from Intergovernmental Organizations	http://www.inchem.org/
National Toxicology Program (NTP) – Chemical Health & Safety Data	http://ntp-server.niehs.nih.gov/Main_Pages/Chem-HS.html; http://www.cdc.gov/niosh/topics/chemical-safety/default.html
Pennsylvania Department of Environmental Protection. Land Recycling Program. Chemical and Physical Properties Database	http://www.dep.state.pa.us/physicalproperties/CASNUM_Search.htm
Safchem Services – SAFCHEM: Database list	http://www.sheilapantry.com/safchem/database.htm
US EPA List of Lists Database	http://130.11.53.73/lol/
WebElements™ Periodic table (professional edition)	http://www.webelements.com/webelements/index.html

- the terminology of the present List, which is the core terminology of TRS
- the terminology of the EDEN-IW Glossary
- pertinent terms from EARTh thesaurus, corresponding to the terms present in Infoterra's EnVoc thesaurus (1997), EEA's GEMET (1999) thesaurus and EEA's Multilingual Glossary (2004)
- synonyms and formulas of pesticides and chemicals in general, obtained from several sources (Table 10.3)
- the terms on the List of Priority substances of the Water Framework Directive
- relevant terms from the text of the same Directive.

In comparison with the TRS, the List encompasses a sub-set of terms of more practical and daily usage, namely a list of basic determinands essential for the query formulation.

Impacted IW Media, Parameters, types of Analytical Fractions, Measurement Units, and so on are presented in the original document (Felluga and Plini, 2004a,b).

The contents of the database of the List

The database of the List contains 268 basic terms/concepts representing the determinands. These terms are identified and selected as "Entry Terms" in British English, Danish, French, and Italian. The List also contains a number of synonyms and non-matching equivalents in the various languages, totalling 463 records, distributed as in Table 10.4.

The origin of concepts/terms

An agreement between NERI and IOW provided the reference List of determinands. Some terms are of very general use, being present in the thesauri and glossaries of the environmental domain; nevertheless, the number of determinands corresponding to pollutants (that are monitored in the water-monitoring stations) is quite high and would be too high to be included as such in a general purpose thesaurus, like EARTh and IOW "Thesaurus de l'eau".

The language coverage

The present status of equivalence between the languages is shown in Table 10.4. Except for the number of synonyms and for the non-matching equivalents, there is complete semantic correspondence between the four languages.

A few terms of German language origin, proposed by IOW, are also present: this is understandable, since they are used for the water bodies in common between France and Germany. They represent an embryonic set of German equivalents.

The content of the record of the List

The structure of the database record includes, for each language, five fields that contain all the basic data for handling and displaying the EDEN-IW TRS terminology:

1. the concept/term ID, obviously unique for all the languages
2. the name of the Type of Determinand
3. the indication of the quality of the term: Entry, Synonym, or Non-matching equivalent (only in English)

Table 10.4. Different numbers of synonyms and non-matching equivalents in the four languages

	English	Danish	French	Italian
Entry terms	268	268	268	268
Having synonyms	160	125	165	166
Non-matching equivalents	35	70	30	29
Total	463	463	463	463

4. the Term, in each language
5. the Specifications.

The latter are only in English, for practical reasons. They contain comments dedicated to the EDEN-IW user, very complex chemical formulas of no practical use for normal users or even equivalents in other languages. The specifications may correspond to the Scope Notes of thesauri or to the Definitions of glossaries.

Tables 10.5 and 10.6 show a sample of the database record: Table 10.5 shows the data used in the EDEN-IW User Agent system, Table 10.6 shows the data presented in the printed version. It can be seen that the Entry Term can correspond to an acronym in one institution and to an expanded form in another.

The Lists

Basically, the results of the work are the following:

For the User Agent of the EDEN-IW ontology system, four abridged monolingual alphabetical lists of determinands; in these lists, only the terms selected as Entry Terms are presented and no synonyms are shown.

For a broader utilisation, a series of complete quadrilingual alphabetical lists was developed. These lists contain the Entry Terms and also, if available, up to five selected synonyms, enclosed in square brackets. The selection of acronyms was restricted to the expanded forms and to the most commonly used chemical synonyms. It should be noted that an individual chemical, like "lindane", can have more than one hundred synonyms, each with a very different frequency of use. In this way, all the terms, Entry Term or Synonym, become entry points, rendering the list of determinand concepts/terms exhaustive. Obviously, there is no perfect matching in the different languages, neither between acronyms and expanded forms nor between the number and nature of synonyms.

Table 10.5. Sample of the database record for the User Agent

	Content of the Field	Example
1	ID	1180
2a	Name of the Type of Determinand, in English	Chemical
2b	Name of the Type of Determinand, in Danish	Kemiske
2c	Name of the Type of Determinand, in French	Chimique
2d	Name of the Type of Determinand, in Italian	Chimico
3a	Quality of the English Term	Entry Term
3b	Quality of the Danish Term	Entry Term
3c	Quality of the French Term	Entry Term
3d	Quality of the Italian Term	Entry Term
4a	English Term	PCB 101
4b	Danish Term	polychlorinated biphenyl 101
4c	French Term	PCB 101
4d	Italian Term	PCB 101
5	Specifications	–

Table 10.6. Sample of the database record for the printed version

	Content of the Field	Example
1	ID	1180
2a	Name of the Type of Determinand, in English	Chemical
2b	Name of the Type of Determinand, in Danish	Kemiske
2c	Name of the Type of Determinand, in French	Chimique
2d	Name of the Type of Determinand, in Italian	Chimico
3a	Quality of the English Term	Entry Term
3b	Quality of the Danish Term	Entry Term
3c	Quality of the French Term	Entry Term
3d	Quality of the Italian Term	Entry Term
4a	English Term	PCB 101 [polychlorinated biphenyl 101]
4b	Danish Term	polychlorinated biphenyl 101 [PCB 101]
4c	French Term	PCB 101 [polychlorinated biphenyl 101]
4d	Italian Term	PCB 101 [bifenile policlorurato 101]
5	Specifications	–

As shown in Table 10.5, an Entry Term can correspond to an acronym in one institution and to an expanded form in another. That would depend on the conventional use in that particular database system.

All the lists were alphabetically ordered by machine, i.e. according to the order of typewritten space, characters, and diacritics of the ASCII system. No other alphabetical ordering system was adopted.

The determinands were clustered by type: Bacteriological, Biochemical, Chemical, and Physical. There were 463 determinands in total, of which 95% belonged to the Chemical type. The Bacteriological (4, synonyms included), Biochemical (10) and Physical (8) determinands are few, and the user will find a separate list of Bacteriological, Biochemical, and Physical determinands, for handy use.

The list of the lists is shown in Table 10.7.

The lists (Tables 10.8 to 10.13) are presented herewith in sample form; the complete lists can be seen in the original document (Felluga and Plini, 2004a,b).

Table 10.7. List of the lists

1. English-ordered Monolingual List of Determinands with Specifications
2. English-ordered Quadrilingual List of Determinands
3. Danish-ordered Quadrilingual List of Determinands
4. French-ordered Quadrilingual List of Determinands
5. Italian-ordered Quadrilingual List of Determinands
6. Determinand Type and English-ordered Quadrilingual List of Bacteriological, Biochemical and Physical Determinands

Table 10.8. English-ordered monolingual List of Determinands with specifications (sample)

Determinand Type	Entry Term (ET)/Synonym (Syn.)	English Term	Specifications	ID
Chemical	ET	% oxygen saturation [oxygen saturation, %]	Percent oxygen of saturated quantity of gaseous oxygen dissolved in water, at the temperature and the atmospheric pressure of the instant of the sampling.	1003
Chemical	Syn.	oxygen saturation, % [% oxygen saturation]	Percent oxygen of saturated quantity of gaseous oxygen dissolved in water, at the temperature and the atmospheric pressure of the instant of the sampling.	1003

10.3 THE TERMINOLOGY REFERENCE SYSTEM

10.3.1 Introduction

The history of Terminology Reference System

In the EDEN-IW project, the work on semantic management through a correct and EDEN-IW-specific terminology included

- development of a reference glossary, partially multilingual
- the collection, terminological handling, and semantic management of the significant terminology of some EDEN-IW databases: actually, the few hundred terms that have been selected by NERI and IOW for the Cross-Mapping Table (see below)
- comparison of the EDEN-IW terminology with the environmental thesaurus of EEA. This comparison was performed in order to reveal interactions (both semantic and terminological) of inland water terminology with the terminology of this thesaurus.

Regarding the EEA thesaurus, it presently includes an obsolete multilingual thesaurus, GEMET, and an updated EEA multilingual glossary. GEMET is no more a reliable source, neither of terms nor of definitions. From the point of view of CNR, GEMET has been replaced by the CNR thesaurus EARTh. The latter contains the terminology of GEMET, duly revised and quality assessed, and presents a much larger term base and a completely new structure.

Originally the EDEN-IW project did not foresee a Terminology Reference System for the domain of Inland Waters or a thesaurus. The terminology required for the EDEN-IW system was supposed to correspond to the needs of the basic queries and to the system of ontology agents developed by other partners, in order to provide the users with appropriate answers. It was agreed that this terminology was in principle multilingual; it had to be

Table 10.9. English-ordered quadrilingual List of Determinands (sample)

Determinand Type	Entry Term (ET)/ Synonym	English Term	Dansk term	Terme français	Termine italiano	ID
Chemical	ET	% oxygen saturation [oxygen saturation, %]	iltmætningsprocent	taux de saturation en oxygène [oxygène, taux de saturation en -]	% di saturazione di ossigeno [ossigeno, % di saturazione di -]	1003
Chemical	ET	1,1,1-trichloroethane [trichloroethane, 1,1,1-]	trichlorethan-1,1,1	trichloréthane-1,1,1	1,1,1-tricloroetano [tricloroetano, 1,1,1-]	1201
Chemical	ET	acenaphthene	acenaphthen	acénaphtène	acenaftene	1161
Chemical	ET	acenaphthylene	acenaphthylen	acénaphtylène	acenaftilene	1176

Table 10.10. Danish-ordered quadrilingual List of Determinands (sample)

Determinand Type	Entry Term (ET)/Synonym	Dansk term	English Term	Terme français	Termine italiano	ID
Chemical	ET	1,2 dichlorethan [dichlorethan 1,2-]	1,2-dichloroethane [dichloroethane, 1,2-]	dichloroéthane-1,2	1,2-dicloroetano [dicloroetano, 1,2-]	1157
Chemical	ET	acenaphthen	acenaphthene	acénaphtène	acenaftene	1161
Chemical	ET	acenaphthylen	acenaphthylene	acénaphtylène	acenaftilene	1176

Table 10.11. French-ordered quadrilingual List of Determinands (sample)

Determinand Type	Entry Term (ET)/Synonym	Terme français	Dansk term	English term	Termine italiano	ID
Chemical	Syn.	1,2,4-Trichlorbenzol [1,2,4-trichlorobenzene; trichlorobenzène 1,2,4-]	–	1,2,4-Trichlorbenzol [1,2,4-trichlorobenzene; trichlorobenzène-1,2,4]	1,2,4-triclorobenzene [triclorobenzene 1,2,4-; triclorobenzolo 1,2,4-]	1249
Chemical	ET	acénaphtène	acenaphthen	acenaphthene	acenaftene	1161
Chemical	ET	acénaphtylène	acenaphthylen	acenaphthylene	acenaftilene	1176

Table 10.12. Italian-ordered quadrilingual List of Determinands (sample)

Determinand Type	Entry Term (ET)/Synonym	Termine Italiano	Dansk term	English term	Terme français	ID
Chemical	ET	% di saturazione di ossigeno [ossigeno, % di saturazione di -]	iltmætningsprocent	% oxygen saturation [oxygen saturation, %]	taux de saturation en oxygène [oxygène, taux de saturation en -]	1003
Chemical	ET	1,1,1-tricloroetano [tricloroetano, 1,1,1-]	trichlorethan-1,1,1	1,1,1-trichloroethane [trichloroethane, 1,1,1-]	trichloréthane-1,1,1	1201
Chemical	ET	acenaftene	acenaphthen	acenaphthene	acénaphtène	1161
Chemical	ET	acenaftilene	acenaphthylen	acenaphthylene	acénaphtylène	1176

Table 10.13. Determinand type- and English-ordered quadrilingual list of bacteriological, biochemical and physical determinands (sample)

Determinand Type	Entry_Syn	English Term	Dansk term	Terme français	Termine italiano	ID
Bacteriological	Entry Term	faecal coliforms	fækale colibakterier	coliformes thermotolérants	coliformi fecali	1062
Bacteriological	Entry Term	faecal streptococci	fækale streptokokker	streptocoques fécaux	streptococchi fecali	1064
…	…	…	…	…	…	…
Biochemical	Entry Term	chlorophyll total [chlorophylls]	total klorofyl	chlorophylle totale [chlorophylls]	clorofilla totale [clorofille]	1025
Biochemical	Entry Term	pheophytin	pheophytin	phéophytine	feofitina	1091
…	…	…	…	…	…	…
Physical	Entry Term	colour	farvetal	couleur mesurée	colore	1007
Physical	Entry Term	conductivity	ledningsevne	conductivité	conduttività	1219

represented in English as a relay language and correspond to other languages to the extent that the equivalents of the English terms were available. A consistent work of translation for the missing equivalents of English terms was not programmed.

A complete multilingual TRS was beyond the scope of the project, also because it would have required the following:

- The EDEN-IW terminology ("language") was known, at least in its broad sense, from the beginning of the project, as regards the extension (number of terms), significance for the ontologies (quality of terms), variety (different types of terms), categorisation (thematic and/ or logical), etc.
- There was an optimal synchronisation between the work of development of ontologies, and the possibility for terms to be collected and translated in the foreseen languages (for reasons inherent to the development of the project, both historical and logistical, the latter were restricted to Danish and French).
- There was continuity of the EDEN-IW endeavour beyond the termination date of EDEN-IW project, a condition that depends on the Information Societies Technology (IST) policy.
- There was a demonstrated requirement ("market") for such a complete system, in both the public and private sector. Such a condition is self-contradictory, unless the potential customer knows exactly what a TRS is, needs specifically an IW-TRS and, last but not least, being familiar with the terminologists in charge to develop the TRS, can rely on the quality of both the work and the final product.

Nonetheless, the work performed during the development of the project for collecting and handling the relevant EDEN-IW terminology and the past experience with multilingual environmental thesauri provided the opportunity to produce a basic TRS for the Inland Water domain, which could be utilised by a large variety of users. TRSs are both useful and required serving as reservoirs of controlled terminology and for translingual access to databases at the same time.

The rationale: The Terminology Reference System in the EDEN-IW context

As for the whole EDEN-IW project, the approach in programming, planning and implementing the TRS was problem-oriented, and the specific application field was the domain of inland waters, which today are precious resources with multiple uses.

The TRS is user-oriented, for the different types of users: decision makers, users of the private sector, public institutions, scientists, professionals, and lay individuals. The final target is to foster and assist the environmental management and the expected benefits are greater environmental sustainability, i.e. optimisation of the human society and its sustainable use of the environment.

The desired assets are both an economy of resources, including the intellectual work needed for the semantic and terminological management, and recovery of the EC investments in the EDEN-IW project.

The means deployed for the implementation of the TRS at programming and implementation levels are both intellectual and technological. Conceptual means for the semantic, terminological, and linguistic management included knowledge organisation, domain terminologies, and translation linguistics. At the technological level, both advanced agent technology and computational linguistics were deployed.

The originality of the EDEN-IW Terminology Reference System

The originality of the EDEN-IW TRS is the presentation of a unique set of linguistic/terminological expressions for each concept/term of the Inland Waters domain, regardless of the preference/non-preference and independently of the language used. In the TRS, terms were traced back to their original multilingual source documents, whose conceptual context was clear, unequivocal, and "unique" and whose semantic management logic was alike, thus proposing a unique multilingual terminology reference system. This system has the form of a sharable multilingual terminology base for the ontology-based queries in the Inland Waters domain, and should help in avoiding redundancy and waste of resources. The unique set of linguistic/terminological expressions means a set of all the linguistic variants belonging to the same concept/term, like the singular and plural forms, the hyphenated forms, and so on.

Issues and problems in implementing a unique Terminology Reference System

Some issues and problems encountered in implementing a unique TRS are listed below:

- Existence of different natural languages and different usage of expressions within each language; the latter is usually subject to small changes, except for the present trend to a global alignment to the American English language, mainly Internet-driven.
- The use in the EU of about 30 languages; in the United States one main language is used, English, with a preference for the Spanish as a second, local language.
- Uncontrolled expansion of the terms used in information technology (IT): technological development, IT systems, databases, etc., in different languages and the uptake from one language into another of these terms.
- Almost uncontrolled expansion of legal and other important languages, in different languages.
- Different definitions of the same concept in different contexts, e.g. in environmental regulations both horizontally, from state to state, region to region, province to province, etc., and vertically, from central to peripheral level.
- Difference in the conceptual approaches, e.g. between librarians and ontologists, between terminologists and linguists.
- Proliferation of glossaries and dictionaries, in different languages, with a fast turnover.
- Slow output by ISO and other standardising bodies.
- Intellectual property rights, copyright, and property aspects.
- Investment costs in semantic, linguistic, translation, and terminology work.

10.3.2 Development of EDEN-IW Terminology Reference System

Materials: Terminology sources

At the start of the EDEN-IW project, a unique, standard, multilingual reference system of terms of the Inland Water domain did not exist. In principle, the "languages" of the IW domain encompassed the academic text of hydrosciences, multilingual technical terminology systems, like "Aqualingua", the internal languages of IW databases and glossaries compiled in different languages or multilingual, like the glossary of UNESCO. A collection and harmonisation of all these terminologies was beyond the scope of the EDEN-IW project. Nevertheless, for the EDEN-IW TRS, in a first step, a quadrilingual Glossary was compiled as a general purpose reference background terminology ("language") of IW domain, in order to be used during the development of the EDEN-IW ontology system and to assist the development of the Basic Multilingual Query Language. This Glossary was mainly based on the quadrilingual IOW "Thesaurus de l'eau", which is an extended set of terms of the multilingual UNESCO International Glossary of Hydrology. The IOW "Thesaurus de l'eau" turned out to be closely connected to the Inland Water databases of France. This thesaurus contained both terms and definitions from the UNESCO Thesaurus and from French sources. For the EDEN-IW project, selected terminology of the IOW Thesaurus was complemented with terms of the Thesaurus EARTh developed by CNR, producing the EDEN-IW Glossary.

At the end of the project, the overall Terminology Reference System of Inland Water – EDEN-IW Terminology Reference System – consists of the following elements:

- the terminology of the Cross-Mapping Table, which is the core terminology of TRS
- the terminology of the EDEN-IW Glossary
- pertinent terms from EARTh thesaurus, corresponding to terms present in Infoterra's EnVoc thesaurus (1997), EEA's GEMET (1999) thesaurus and EEA's Multilingual Glossary (2004)
- for synonyms and formulas of pesticides and chemicals in general, the Compendium of Pesticide Common Names, http://www.alanwood.net/pesticides/index.html
- the terms of the list of the Priority substances of the Water Framework Directive
- relevant terms from the text of the same Directive
- the keywords of EEA Waterbase, 13 May 2004 edition (updating and some changes are obviously ongoing).

Methods and procedure

A stepwise approach was taken and the different steps are listed below.

First step

- interaction with IOW Thesaurus management personnel
- quality control of the English equivalents of the IOW Thesaurus
- comparative analysis of EARTh and IOW thesauri
- merging, concept-based
- orthographic checking and normalisation
- collection of definitions and their sources
- control of semantic coherence between terms and definitions
- control of synonyms, duplicated definitions, and crossed references
- orthographic control.

FINAL step

- definition of the Data Dictionary of TRS database
- selection of the base terminology material: EDEN-IW Cross-Mapping table terminology; EDEN-IW Glossary; the compendium of Pesticide Common Names for the synonyms of pesticides; terminology of the Water Resources framework Directive (WFD); terminology of EEA Waterbase
- interaction with both NERI and IOW for the Danish and French equivalents, respectively
- identification, selection, and preparation for merging of terms from both sources
- merging, concept-based
- orthographic checking and normalisation
- collection of definitions and their sources
- control of semantic coherence between terms and definitions
- control of synonyms, duplicated definitions, and crossed references
- translation into Italian
- orthographic control
- reporting and editing.

10.3.3 Results

The Terminology Reference System database content

The EDEN-IW TRS database contains 2383 terms/concepts in UK English, with partial equivalence in Danish, French, and Italian; it contains also 1083 synonyms in English, totalling 3466 terms in the reference language.

The origin of concepts/terms

The TRS terms originate from 17 different sources, one of them dating back to 1983 (the MDS, Multilingual Descriptor System of the European Commission). The terminology of the different sources obviously overlapped to some extent. The number of terms referred to their original sources is detailed in the Tables 10.14 and 10.15.

It is interesting to note that starting from the MDS of 1983, there is continuity in the utilisation of a number of the very same terms/concepts in different environmental documents used by different institutions. It also appears that, even if not explicit, there is the need of a *de facto* shared terminology for the handling of data and metadata in different databases.

The language coverage

In the EDEN-IW project, there was the need of a language ensuring 100% coverage, essential for the navigation between the different databases. The equivalents in other languages cited in the table were added according to their availability. The number of equivalents is also a consequence of the different size of the databases.

Table 10.14. Origin of Terminology Reference System terms/concepts, by year and original source

No.	Term Source in EDEN-IW	Basic Doc.	Doc. Year	By Inst.	No. of Entry Terms/Concepts	% (round.)	Total No. of Terms/ Concepts including Synonyms
01	Cross-Mapping Table Determinand Folder	MDS	1983 – > 1989	EU	503*	21	511*
02	Glossary + EARTh 2003	U84	1984	UN	1	0	1
03	Glossary + EARTh 2003	BDS	1989	IT	50	2	50
04	Glossary + EARTh 2003	U97	1990– > 1997	UN	7	0	8
05	Glossary + EARTh 2003	TIA	1990– > 1994	IT	107	4	108
06	Glossary + EARTh 2003	TMA	1990– > 1995	SP	66	3	67
07	Glossary + EARTh 2003	CLS	1995	EU	44	2	44
08	Glossary + EARTh 2003	UMW	1995	DE	85	4	85
09	Glossary + EARTh 2003	EVO	1995	EU	**	–	**
10	Glossary + EARTh 2003	LEX	1995	FR	200	8	200
11	Glossary + EARTh 2003	EEA	1997	EU	1	0	1
12	Glossary IOW Thesaurus (incl. UNESCO)	UNE	1999	UN	457	19	457
13	Glossary IOW Thesaurus (incl. UNESCO)	IOW	2000	FR	191	8	191
14	Cross Mapping Table_WFD Priority Substances List	WFD	2003	EU	49	2	78
15	Cross-Mapping Table Determinand Folder	CMT	2003	EU	294	12*	1336
16	Glossary + EARTh 2003	EAR	2004	IT	28	1	29
17	EEA Waterbase 2004	WAT	2004	EU	300***	13	300***
							Synonyms 1083
					Total: 2383	Total: 100	Total: 3466

* Specific contribution, with instrumental Terms, like Units, etc.

** Eurovoc Multilingual Thesaurus of the European Parliament. The terms of this thesaurus represented in the Terminology Reference System were original of other documents.

*** Terms can be coinciding with other Waterbase terms and/or with terms of other origin, if contexts are different.

(Continued)

Table 10.14. (Continued)

Legend of columns

No.	Sequential number by year
Term Source in EDEN-IW	Glossary or Cross-Mapping Table sources
Basic Doc.	Basic document: thesaurus, lexicon, glossary, list, text
Doc. Year	Basic year of the document
By Inst.	Symbol of the Institution; UN (United Nations system); EU (European Union); the other are national contributions
No. of Entry Terms/Concepts % (round.)	Percentage, rounded
Total no. of Terms/Concepts including Synonyms	ditto

Legend of symbols – Basic Document

No.	Symbol	Basic Document
01	MDS	Multilingual Descriptor System (1983).
02	U84	Infoterra Thesaurus (1984).
03	BDS	Bilingual Descriptor System (1989).
04	U97	Infoterra Thesaurus 1990 and EnVoc 1997.
05	TIA	Thesaurus Italiano per l'Ambiente: covers both the trilingual (1990) and the quadrilingual Thesaurus for the Environment (1994).
06	TMA	Tesauro de Medio Ambiente (monolingual, 1990) and its quadrilingual version (1995).
07	CLS	Classification Scheme for the Multilingual Environmental Thesaurus (1995).
08	UMW	A selection of 2032 terms from the about 8000 terms of Umwelt Thesaurus (1995).
09	EVO	Eurovoc Thesaurus of the European Parliament (1995).
10	LEX	Lexique environnement, Paris (1995).
11	EEA	DPSIR Scheme, EEA (1997).
12	UNE	UNESCO Multilingual Glossary of Hydrology (1999).
13	IOW	Thesaurus de l'Eau, multilingual (2000).
14	WFD	Water Resources Framework Directive (updated 2003).
15	CMT	Cross-Mapping Table of EDEN-IW: the total number is largely expanded by the synonyms of pesticides.
16	EAR	EARTh thesaurus of CNR (2004).
17	WAT	EEA Waterbase (2004-05-13).

Table 10.15. Origin of Terminology Reference System terms/concepts, by level of contributing institution, from international to national

No.	Basic Doc.	Doc. Year	By Inst.	No. of Entry Terms/Concepts	% (round.)	Total No. of Terms/Concepts including Synonyms
02	U84	1984	UN	1	0	1
04	U97	1990 – > 1997	UN	7	0	8
12	UNE	1999	UN	457	19	457
01	MDS	1983 – > 1989	EU	503	21	511
07	CLS	1995	EU	44	2	44
09	EVO	1995	EU	**	–	**
11	EEA	1997	EU	1	0	1
14	WFD	2003	EU	49	2	78
15	CMT	2003	EU	294	12**	1336
17	WAT	2004	EU	300	13	300
03	BDS	1989	IT	50	2	50
05	TIA	1990 – > 1994	IT	107	4	108
16	EAR	2004	IT	28	1	29
06	TMA	1990 – > 1995	SP	66	3	67
08	UMW	1995	DE	85	4	85
10	LEX	1995	FR	200	8	200
13	IOW	2000	FR	191	8	191
						Synonyms 1083
				Total: 2383	Total: 100	Total: 3466

Legend of symbols – Institution

Symbol	Institution
UN	United Nations System: UNEP, UN Environment Programme, Nairobi; UNESCO, UN Education, Scientific Commission, Paris.
EU	European Union: European Commission, Bruxelles; European Environment Agency-Task Force, Bruxelles; European Environment Agency, Copenhagen; European Parliament, Strasburg.
IT	CNR, Consiglio Nazionale delle Ricerche, Rome.
SP	MMA, Ministerio de Medio Ambiente, formerly MOPTMA, Ministerio de Obras Publicas, Transportes y Medio Ambiente, Madrid.
DE	UBA, Umweltbundesamt, Berlin.
FR	Ministère de l'environnement, Paris; IOW, Institute Of Water, Paris – Limoges.

Table 10.16. Status of equivalence between the languages

Language	No. of Entry Terms/Concepts	Total No. of Synonyms in the Complete Alphabetical List	Total No. of Terms/Concepts, including Synonyms
English-UK	2.383	1.083	3.466
Danish equivalents	2.383	124	2.507
French equivalents	2.383	394	2.777
Italian equivalents	2.383	279	2.662
	Total: 5.889	Total: 1.880	11.412

The content of the Terminology Reference System record

Table 10.17. Sample of the Terminology Reference System database record

Eng. UK Term	temperature
Eng. UK Synonyms	–
Definition	Temperature of "MEDIUM" at the moment a sample is taken.
Source 1	EDEN-IW CMT [See separate List]
Source 2	–
Source 3	–
Source 4	–
Danish	temperatur
Danish Synonyms	–
French	température
French Synonyms	–
Italian	temperatura
Italian Synonyms	–
[Term_Function Entry Term (Synonyms will be in italics)]	
Orig. Term Inst.	EC-ENREP – > CNR
Orig. Term Doc.	MDS – > BDS
Orig. Term Year	1983 – > 1989
CMT_YN	Y
ECB_YN	ECB -
WFD-P_YN	N
EWN	–
ISONamPes_YN	ISO -
EWN	–
Term Type	–

Table 10.17. (Continued)
Legend

Eng. UK Term	Ditto
Eng. UK Synonyms	Ditto
Definition	Ditto
Source 1	Source of Definition (1). See separate List
Source 2	Source of Definition (2). See separate List
Source 3	Source of Definition (3). See separate List
Source 4	Source of Definition (4). See separate List
Eng. US Term	Ditto
Eng. US Synonyms	Ditto
Danish	Danish equivalent
Danish Synonyms	Ditto
French	French equivalent
French Synonyms	Ditto
Term_Function	Entry Term (Synonyms will be in italics)
Orig. Term Inst.	Institution providing the first English-containing multilingual Document presenting the Term for the first time
Orig. Term Doc.	Document, thesaurus or lexicon or glossary, presenting the Term. See Legenda of Symbols, above
Orig. Term Year	Year of publication of the Document
CMT_YN	Presence in the EDEN-IW Cross-Mapping Table.
	Instances
	N
	N_EARTh
	N_EARTh + IOW
	N_IOW < UNESCO
	N_New Term
	N_WFD Text
	Y_Analytical Fraction
	Y_Determinand
	Y_Determ. + Unit Comp.
	Y_Medium
	Y_Methods
	Y_Subjects
	Y_Units
	Y_Value Attribute
	Y_WFD Prior. Subst.
ECB_YN	Presence in the European Chemicals Bureau List. See: ecb.jrc.it/
ISONamPes_YN	Presence in the Index of Common Names. See, e.g. http://www.hclrss.demon.co.uk/
WFD-P_YN	Presence in the Water Resources Framework Directive – Priority List. See: http://europa.eu.int/comm/environment/water/water-framework/index_en.html
EWN	Indication in the Euro-Waternet classification of inland water resources into three main categories: rivers and canals, lakes and reservoirs, groundwater.

Table 10.17. (Continued)

	WFD_PS = Water Resources Framework Directive – Priority List See: http://www.oieau.fr/ciedd/contributions/at1/contribution/eea.htm	
	Instances and acronyms: IW bodies where determinands are determined: Groundwater Lake: lakes and reservoirs River: rivers and canals The data collected for Eurowaternet (and therefore candidate to be in Waterbase) are for a limited set of determinands: R = determinand requested for rivers stations. L = determinand requested for lakes stations. G = determinand requested for groundwater stations.	
	DS = the determinand is part of the substances list of the Dangerous Substances directive **DW** = the determinand is part of the substances list of the Drinking Water directive **WFD_PS** = the determinand is part of the priority substances list of the WFD (Annex 10) **WFD_PSR** = the determinand is part of the priority substances under review of the WFD (sublist of WFD_PS) **WFD_PHS** = the determinand is part of the priority hazardous substances list of the WFD (sublist of WFD_PS)	
Term Type	Instances	Example
	–	lead
	Acronym	PAH
	Acronym – Abbreviation	TRI-6
	CAS Number	577-11-7
	Convention Name	Ramsar Convention
	ECB Name	ethylene dichloride
	EINECS Number	202-049-5
	Formula (Molecular)	Sb
	Formula – Abbreviation	T2
	Geogr. Name	Aral sea
	Micro-organism Name	Escherichia coli
	Symbol	m^2

Table 10.18. The Alphabetical Lists

1	Complete English-UK-based Alphabetical List
2	English-ordered Quadrilingual List
3	Danish-ordered Quadrilingual List
4	French-ordered Quadrilingual List
5	Italian-ordered Quadrilingual List
6	List of Definition sources

Table 10.19. Complete English-UK-based Alphabetical List (sample)

English UK Entry Terms and *Synonyms*	English UK Entry Terms and *Synonyms*
%	per cent; percentage
...	...
HCCH	lindane
...	...
urban water	–
V (chemical element)	vanadium
V (electricity unit)	volt
Vanadium	V (chemical element)
...	...
V-notch weir	–
Volt	V (electricity unit)
...	...
vulnerable ecosystem	–
Water	–
water (substance)	–
water acidification	–

Table 10.20. English-ordered Quadrilingual List (sample)

English UK ET	Danish Term	French Term	Italian Term
Lindane	lindan	lindane	lindano
...	...	...	...
low flow	strømning	étiage	magra

The Lists

The English-based lists are presented herewith in sample form in the Tables 10.16 to 10.21. The complete lists are presented in the original document (Felluga and Plini, 2004a,b).

The Danish-ordered, French-ordered, and Italian-ordered Quadrilingual Lists can be seen in the original document (Felluga and Plini, 2004a,b). A screenshot of a final EDEN-IW TRS record is shown in Figure 10.1.

10.4 DISCUSSIONS AND CONCLUSIONS

The issues and problems encountered in implementing the present Quadrilingual List of determinands are analogous to those faced when developing a unique TRS for EDEN-IW. The following are the significant points:

- The uncontrollable expansion of the IT language in its various aspects – technological development, IT systems, databases, etc. in different languages.

Table 10.21. List of definitions sources (sample)

Abbrev.	Author	Title	URL	Publication	Place	Year
ABDN	—	—	http://www.abdn.ac.uk/~pol028/sources/europe.htm	—	—	—
ACRA	—	—	http://www.acra.it/acrateca/glosster.htm	—	—	—
ADMIN	—	—	http://www.admin.ch/buwal/i/themen/umwelt/vielfalt/ik14u05.pdf	—	—	—
AERG	—	—	http://aerg.canberra.edu.au/pub/aerg/davey/iucn...	—	—	—
AGBUS	—	—	http://agbusmgt.ag.ohio-state.edu/ae601/glossary/glosse.htm	—	—	—
AGP	—	—	http://frost.ca.uky.edu/agripedia/tagrimai.htm	—	—	—
AGRENV	—	—	http://www.agrenv.mcgill.ca/EXTENSION/ECOMUSE/E...	—	—	—
AGRIC	—	—	http://www.agric.gov.ab.ca/agdex/500/4200009.html	—	—	—

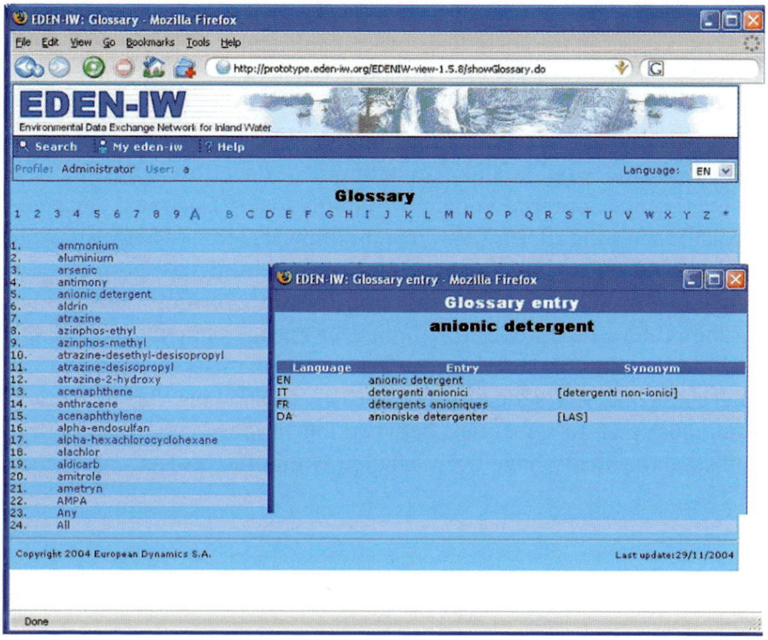

Fig. 10.1. Screenshot of a final EDEN-IW TRS record.

- The uncontrollable expansion of specific languages of relevance for the environment, like the legal, the information, the technical, in different languages.
- The different definitions of the same concept in different contexts, e.g. in environmental regulations both horizontally, from state to state, region to region, province to province, etc., and vertically, from central to peripheral level.
- The proliferation of glossaries and dictionaries, in different languages, with a fast turnover.
- The slow output by ISO and other standardising bodies.
- The investment costs in semantic, linguistic, and terminology work.

The present List, which is restricted to a rather small and uniform domain, shows the following:

- A multilingual list of determinand is feasible, provided strict rules of multilingual terminology handling of concepts/terms are followed.
- A consensus among water institutions of different countries, speaking different languages, can be reached and might represent a challenge towards the harmonisation of the procedures for the control of the IW quality across the EU member countries. This would solve the problems due to different methodologies, e.g. different measurement and/or monitoring methods, different instrumental setup, and approaches, different regulatory languages, etc.

- From the terminological standpoint, a mandatory step towards this consensus would be the harmonisation of the technical, procedural, and regulatory terminologies: this would solve the problems of the conceptual differences, i.e. of different meaning for a same term.

It would be worthwhile to extend the multilingual coverage of the List to other European languages; this would require a continuity of the effort started in the EDEN-IW Project. Potential sources of terminologies of international or national databases that may become associated with the EDEN-IW project are presented in Table 10.22.

There are various approaches in managing the multilingual terminology, and they include the application of criteria for choosing (or not choosing) a working language, for handling the preference criteria, for forcing or not the equivalence, for handling the borrowed terms, for ordering the terms, for presenting the lists, and so on.

In the present work, the added value consists in showing the differences in term representation, more than in semantics, between the various languages: these differences must be handled and actually have been handled within the project.

Table 10.22. Potential sources of additional terminology to the EDEN-IW determinand list

No.	Organisation	Source Lang.	Website	Database URL
01	European Topic Centre on Water	EN	http://water.eionet.eu.int/Databases	
02	SEPA, Scottish Environment Protection Agency	EN	http://www.sepa.org.uk/	http://www.sepa.org.uk/data/hm/hm.asp
03	Environment Agency for England and Wales	EN	http://www.environment-agency.gov.uk/	http://216.31.193.171/asp/1_map.asp
04	Flemish Environment Agency	NL	http://www2.vmm.be/servlet/be.coi.gw.servlet.MainServlet/id1047563274543/standard/?toDo=open&id=51	http://www2.vmm.be/servlet/be.coi.gw.servlet.MainServlet/standard?toDo=open&id=1269&&
05	Portuguese Institute for Water	PT	http://www.inag.pt	http://snirh.inag.pt/snirh/dados_sintese/main_nav_fr.html
06	Federal Environment Agency, Osterreichische Umweltbundesamt	GE	http://www.ubavie.gv.at/	http://www2.ubavie.gv.at/8412/cgi-bin/q001/q001.html

For the presentation of the lists, the approach adopted in the List is somewhat original: it uses a unique field for the presentation of both the Entry Terms and the selected Synonyms, easing the implementation of the list in the EDEN-IW system.

The approaches adopted to handle the determinand multilingual terminology in the List have been applied also in the EDEN-IW TRS. The present effort was made with the following aims:

- To optimise the use of resources.
- To demonstrate that the concepts belonging to the environmental domain can be assumed to be the same across the present time and space as elements of a universal environmental knowledge.
- To demonstrate that this environmental knowledge – expressed in the different languages and collected in multilingual documents – could act as a common, globally shared language.

The extension of EDEN-IW TRS is, in principle, unlimited: its implementation will depend on the availability of time and resources. The extension could include the following terminologies that can be relevant for EDEN-IW:

- Waterbase terminology.
- Terminologies of international or national databases that may become associated to the EDEN-IW project.

The EDEN-IW TRS provides an example, albeit in prototypical form, of multilingual terminology harmonisation that could foster the development of a common standard terminology for the inland water domain.

Chapter 11

AGENT COMMUNICATION VISUALISING TECHNOLOGIES

J. Fowler, A. R. Cassandra, G. E. Pitts and R. J. Lozado

11.1 INTRODUCTION

The design of the EDEN-IW System envisioned the development of a tool for the graphic display of the communications between EDEN-IW agents, as it was important to demonstrate the performance of the EDEN-IW System to end-users and sponsoring partners and agencies, as well as for debugging the agent communications during system development. The Agent Communication Visualiser (ACV) was implemented specifically to support the agent technology utilised in the EDEN-IW System. Despite the ACV's focused initial use, it was designed with more general usage in mind. It is anticipated that this tool might be useful for displaying communications between potentially distributed agents in other Internet Protocol-based agent systems. This should be possible with only minor modifications (in many cases, simply modifying the configuration will suffice). The source code for ACV is publicly available on the Web along with demonstrations of sample configurations.

11.1.1 History of the EDEN Monitor

The successes and limitations of the EDEN project, which was the precursor to EDEN-IW, have been described in Chapter 4. One component of that project that was significant for the way in which it helped promote audience understanding of system demonstrations was the InfoSleuth Monitor Agent and the applet it used to display messages in an agent conversation. Because of the significant benefit provided by the monitor for communicating the basic performance of the EDEN system to users, visualisation of agent messages was seen as a similarly significant need in the EDEN-IW project.

There were two versions of the InfoSleuth monitor, each of whose characteristics elicited different impressions in viewers. The first version displayed each agent as a named box in a column of boxes. This monitor used a yellow arrow emanating from a box to indicate a query, and a red arrow to indicate a reply. Each arrow was flashed three times; since there was no synchrony in displaying the sending and receipt of messages, these "flashing lights" gave the false impression that a complex conversation was occurring between the agents, when in fact only a simple query and response had occurred.

The second version of the monitor again used a column of named boxes for the agents. However, it then traced messages along the paths of a "circuit diagram" that showed all the possible paths between agents. This version eliminated the false impression of complexity produced by the first, but there remained the significant issue of asynchronous display of essentially synchronous events.

11.1.2 The Jade Sniffer Agent

The EDEN-IW agent software was constructed using the Java Agent Development Environment (JADE). JADE provides a so-called Sniffer Agent that subscribes to the message traffic of selected agents in the system and displays the messages that it detects. The Sniffer display loosely resembles a UML process diagram. It consists of a series of vertical lines representing individual agents, with messages indicated as horizontal arrows passing from the line representing the sender to the line representing the receiver. Messages are laid out from top to bottom of the display in temporal order of their receipt by the Sniffer – not in the order of their actual occurrence. Thus, the Sniffer exhibits the same problems of asynchronous display of messages presented by the EDEN monitor.

11.1.3 Perceived Challenges in Visualising Agent Communication

Both of the agent communication visualisation tools just discussed had several unresolved limitations that the EDEN-IW development team wished to address in a new ACV. The first challenge to be overcome was the ability to display more than a small handful of agents in a limited screen space. The nature of the EDEN monitor's "circuit diagram" demanded screen real estate (i.e. space) linear in the number of agents in the vertical dimension, and related to the square of the number of agents in the horizontal dimension. The Sniffer requires screen real estate linear in the number of agents in the horizontal dimension and linear in the number of messages in the vertical dimension.

A second related challenge was to permit a user as much flexibility as possible in the choice of display attributes. A design requirement was to be able to map agent locations onto an arbitrary background image, which might be a geographic map or an image depicting logical relationships between agents. A challenge deemed to be beyond the scope of the design requirements was to create a graphic editing environment that permitted developers to modify and save display configurations.

Finally, there is the significant challenge of imposing some semantic order on an asynchronously detected message stream in order to make the displayed result more meaningful to a viewer. In both the EDEN monitor and the JADE Sniffer Agent, there is no synchrony in the display of messages. The perception problems this created are several. Foremost, of course, the actual time required for the transmission, computation, and reply of a query can be far too brief for a human to perceive, so messages travelling between agents must necessarily be depicted in an artificially long duration. This creates a second visualisation problem, because the monitor frequently receives reply messages during the interval in which their corresponding query is being displayed. Because there are no semantic constraints placed on message ordering, the monitor then immediately begins to display the reply, with the unfortunate result that the viewer sees a reply being sent before the query that prompted it has been received. This asynchronous display of what is essentially a synchronous directed graph of a conversation between many agents leads to many similar logical inconsistencies in the attempt to display agent message traffic in near real time.

11.2 GOALS OF THE AGENT COMMUNICATION VISUALISER

The challenges just cited became the main goals of ACV development, and were represented explicitly in the ACV requirements document. The high-level goal of the project was to provide a tool for viewing agent communications that would facilitate user understanding of EDEN-IW system performance. A secondary goal was to provide a debugging aid for the final stages of development of the EDEN-IW System. Our design thus emphasised the following attributes.

11.2.1 Flexibility of Display

Two different aspects of flexibility in display were envisioned in the ACV development goals: the ability of a system designer to prepare a particular visualiser configuration for use in an application or demonstration, and the ability of a particular application user to modify a pre-configured visualiser for personal preferences.

11.2.2 Near–Real-Time Portrayal of Semantic Dependencies

Although it cannot be guaranteed that an observer of a concurrent system will see the events in that system in their actual order of occurrence, there are certain axioms of communication behaviour that can be used to impose additional logical ordering on the observed event stream. Agent conversations are generally characterised by typed messages; the simplest non-trivial conversation is the remote procedure call (RPC), which involves a query and a reply. Given information about the types of messages in an RPC, one can impose the semantic constraint that a reply from agent B to agent A must have been preceded by a query from agent A to agent B, providing an ordering mechanism to support this and other behaviours was a goal of the ACV.

11.2.3 Simplicity of Adaptation

Although the ACV was conceived for use in the EDEN-IW project, it was recognised that the visualisation functionality of the ACV might be useful in other agent systems or in altogether different settings. Hence, consideration was given to providing simple interfaces through which to communicate message streams to the ACV.

11.2.4 Utility for Debugging

Due to the speed and inherently distributed nature of computer inter-process communications, it is not feasible to use any visualisation tool for debugging of agent systems in real time. However, experience with EDEN-IW showed that visual cues could be useful to developers trying to understand communication behaviour in an agent system. In addition, the ability of the tool to create a log of messages allows post-mortem inspection of communication behaviours.

11.3 THE ACV PROJECT

The design of the ACV was motivated by the inflexibility of the two example systems discussed above. The watchword of the development process was to make possible many visible modifications to system behaviour without the use of programming, and to provide a structure within which further programming enhancements were facilitated by modularity of design.

11.3.1 Design

Following current programming practice, the ACV was designed with a *Model-View-Controller* architecture. The *Model* module simply expresses the relationships between agents and the messages they send. The *Controller* module provides the handling mechanism for messages between agents by means of a *ConversationHandler*, subclasses of which impose different kinds of semantics on message ordering. Most of the Java code of the ACV is to be found in the *View* module. This module receives events from the *ConversationHandler* by means of the *AgentEvent* interface, and then displays them as appropriate. In addition to managing message events, the *AgentEvent* interface is designed to support changes in agent state or location, but a module for the implementation of agent state events has not been developed.

11.3.2 Development

Programmer responsibility for the development of the ACV was divided across the *AgentEvent* interface used by the *ConversationHandler* to communicate with the graphic display engine. These two divisions are discussed below.

Graphic display engine
In the development of the graphic display, a prime goal was that all visual characteristics of agent and message appearances must be configurable by means of a standard Java

properties file. Configurable characteristics include the background image and possible presence of multiple viewing frames; message shape, colour, and size; agent icon, size, and visibility; colour and relative location of agent labels and statistics windows; and shapes of individual message paths and traversal speeds.

The use of a hierarchical inheritance model for the characteristics of agents provides great configuration flexibility by allowing attributes to be configured at several levels. First, in the absence of a configuration, the ACV maintains a set of default attribute values for all agents and messages. Next, a global default can be configured for all agents. Then, agent attributes can be set by agent type. Finally, individual agents can be configured specially. This allows simple configurations with a few types of agents to be easily managed without sacrificing the ability to develop a complex interface requiring many specially configured agents.

Provision is also made for defining the paths of messages between agents explicitly, but a standard set of path layouts is provided, with simple straight lines as the default. This approach makes simple configurations easy and complex configurations possible. The custom message paths are particularly useful when physical limits or important topological information needs to be conveyed.

Regardless of the available screen real estate, complex configurations can still result in significant clutter when there are many interacting agents. To allow focusing on subsets of the full agent system (either for demonstration or debugging) agents and/or their messages can be visually hidden.

Important for debugging are tracking statistics about agent interactions. For example, the number of messages exchanged per agent can often lead to understanding trouble spots in the communications. The ACV tracks the statistics and optionally allows them to be shown near the agent on the screen. The location, relative to the agent, of where the statistics appear is configurable.

In order to manage large configurations, the ACV provides the ability to zoom in and out of the display, as well as the ability to simultaneously display multiple windows, each with its own view of the full system.

Communication semantics

The semantics of message ordering is enforced by the *ConversationHandler* interface, which manages objects implementing the *OrderedMessage* interface. The *BaseMessage* class, which implements the *OrderedMessage*, also extends the graphic display engine's *AgentMessage* in order to deliver events for graphic display. The basic implementation of the *ConversationHandler* is the *SimpleConversationHandler*, which simply emits a message as soon as the event handler detects it.

The first imposition of semantics on the message stream is the *RPCConversation Handler*, which enforces the semantics of the RPC. There are two kinds of assumed messages: query and reply. A reply from agent B to agent A will be displayed after the query from agent A to agent B even if it is observed first. In addition, we make a "query-chaining" assumption that if we see agent B send a query message to agent C before replying to agent A, then the message from B to C was a response to the stimulus of the query message from A to B (that is to say, we assume that it would not have been sent if agent B had not received the query from A), so we do not display the reply to agent A until we have also seen the reply from agent C to agent B. The query-chaining assumption is invoked recursively when other messages meet the prescribed conditions.

The *JadeConversationHandler* is a subclass of the *RPCConversationHandler* that enforces the logical semantics of JADE as used in EDEN-IW. In addition to the two kinds of messages for RPC, we also assume that an agent that receives a query immediately sends a "confirmation" message before doing its computation and then sending a reply. The *RPCConversationHandler*'s assumptions about query chaining are enforced by inheritance and extended to the chaining of "confirmation" messages.

Another handler, the *GlobalConversationHandler*, uses a single global counter with a unique value in each message; messages are displayed in increasing order of the global counter. The *SimpleConversationHandler* is actually a degenerate subclass of the *Global ConversationHandler* in which the global counter is maintained by the ACV and it is simply incremented with each message detected by the event handler. This ordering is nothing more than the order of receipt by the ACV. Therefore, the *SimpleConversation-Handler* just reproduces the semantically inconsistent message display behaviour of the EDEN monitor and the JADE Sniffer.

One more order model was contemplated, but was not implemented because it was beyond the needs of the EDEN-IW project. This was the so-called Lamport order model, so named for Leslie Lamport's logical ordering for systems in which messages that are monitored contain a sender ordering number and a receiver ordering number.

11.3.3 Project Results

A reasonably ambitious design specification was adopted, and its goals were not only met, but to some extent exceeded. As desired, the ACV can display EDEN-IW messages that it has received across a network. However, it can as well play back messages read from a log file. In addition, it can impose not only a JADE-based semantic ordering, but also RPC ordering as well as sequential ordering. The ACV also has a "VCR" control panel that allows the network input to be stopped and then replayed or stepped through the recorded agent conversation. The most impressive accomplishment, however, is the flexibility of configuration. The initial design specification simply desired start-up time configurability. The version of the ACV released for the final EDEN-IW demonstration included not only full configuration from a configuration properties file, but also an editing mode that allowed dynamic modification of all aspects of agent and message appearance as well as saving the edited configuration. The aspect in which the design specification was incomplete was support for internationalisation. The ACV supports translation resources for menu items of the display, and the current release includes resource bundles for Danish, French, and Texan.

11.3.4 Alternative Implementations

The existing corpus of code for the ACV was not designed with a specific communications architecture in mind; however, to satisfy the EDEN-IW project requirement for a Web-accessible visualiser, the implementation that exists runs as a Java applet with a simple connector for reading message streams from a network socket. A message consists of a simple ASCII byte stream. The messages are delimited by an end-of-line character and fields are separated by tabs. An obvious enhancement would be to define a formal protocol that uses XML to deliver structured messages to the ACV.

Communication from the JADE Sniffer was accomplished by a minor modification to `Sniffer.java` to allow it to emit an appropriate message on a network socket. This had the benefit of a very simple implementation, but it is not the only way in which one might consider generating a message stream for display. Other agent systems may present alternative possibilities.

An alternative implementation of the ACV for JADE might embed the ACV directly into the Sniffer tool, using Java communication methods directly. This would have the advantage of exploiting the ACV's abstract message type to allow structured message objects.

Another approach to communicating with the ACV would be to wrap the ACV as a service using the Simple Object Access Protocol (SOAP).

11.4 THE ACV IN USE

A screen capture of the ACV is shown in Figure 11.1. Visible is the main display, with a map of Europe in the background and agents distributed across the map, and paths and messages travelling between agents. In the other windows, the agent legend and the message legend are visible, along with the terminal Java window showing the list of messages being displayed.

Figure 11.1 shows the ACV running as an applet on Mac OS X during a query session of EDEN-IW. The positions of the agents and services are identified on a catchment-basin map of Europe, along with explanatory legends that identify the types of EDEN-IW agents and the messages that pass between them. Messages are displayed as different-coloured shapes that travel along the paths between agents. Legends for agent types and message types are displayed, as well as dynamic per-agent statistics showing messages sent and received. Another window displays the log of messages to and from individual agents.

In Figure 11.2, the ACV, running in the Mozilla Web browser, displays a map of Europe in the background, with the visible agents located in their approximate real physical locations. A red diamond indicates the return of a reply from the Directory Agent to a User Agent located in Greece. Meanwhile, the Task Agent has just issued a request, indicated by the star, to the UK Resources Agent. The IOW Resource Agent is replying to the Task Agent at the same time. The Run menu is selected, showing some of the possible ways of controlling the display of the messages, including pausing the display, stepping one-by-one through the message log, or saving the log for later examination or

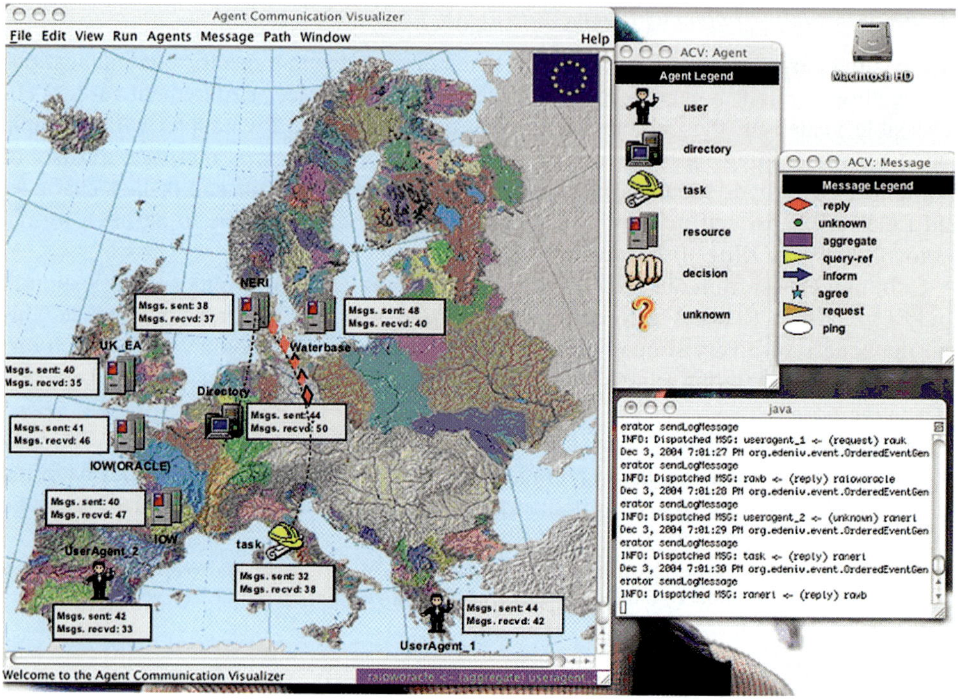

Fig. 11.1. The ACV displays a reply message from NERI to the task agent.

replay. The ACV display background can be altered as well, so that the message traffic pattern between a certain groups of agents can be viewed in more than one way. For example, while Figure 11.2 shows the agents displayed on a map, simple changes to the configuration file make it possible to show the same agents displayed on a graphic image that shows the logical relationship of the EDEN-IW agents, with User agents appearing on the left-hand side of the screen, the active agents of the JADE middle tier in a "cloud" at the centre, and the resource agents from various water quality databases appearing on the right.

Figure 11.3 shows the "cloud" demo, a Web page constructed from a PowerPoint image. The running applet displays a sequence of red diamonds to indicate a reply from a broker or directory agent to one of the user agents. At the same time, the status window at the bottom indicates that a user agent has issued a query to the query agent. Numerous messages have been sent and received by all agents, which is indicated by the statistics boxes displayed below each agent.

The event model of the ACV refers to the source of agent events and messages. There are three current event models: "random", "logfile", and "network". The default, which is useful for demonstration, is "random". Here messages and new agents are created at random. When the event model is "logfile", messages are read from a file. These are

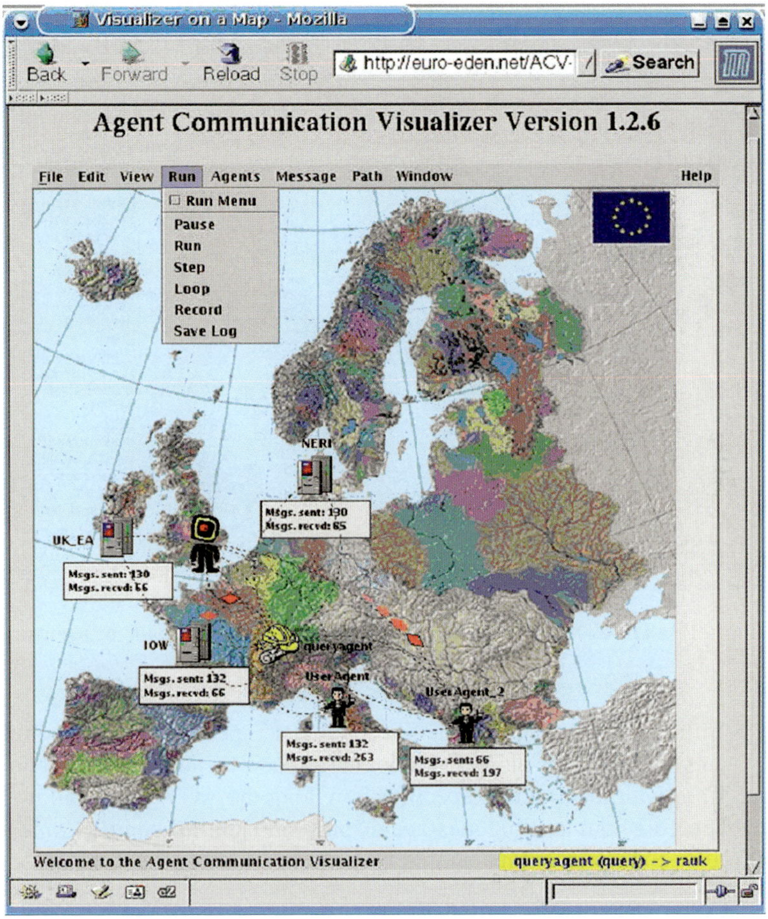

Fig. 11.2. The EDEN-IW agents and the control menu.

delimited one message to a line. This is the same format in which messages are saved to files. The log file from which to read the messages is specified in the configuration file. The final event model is the "network" model, in which messages are read from a network socket. As with the "logfile" model, messages are delimited one message to a line. The socket on which to listen for messages is specified in the configuration file. The default port is 1997. To run in network mode from a Web page, the user's Java security policy must be set to permit the applet to accept connections on the network socket.

11.4.1 System Configuration

Some important properties of the ACV must be specified in the configuration properties file. The "order.model" identifies whether the ordering semantics of a particular kind of

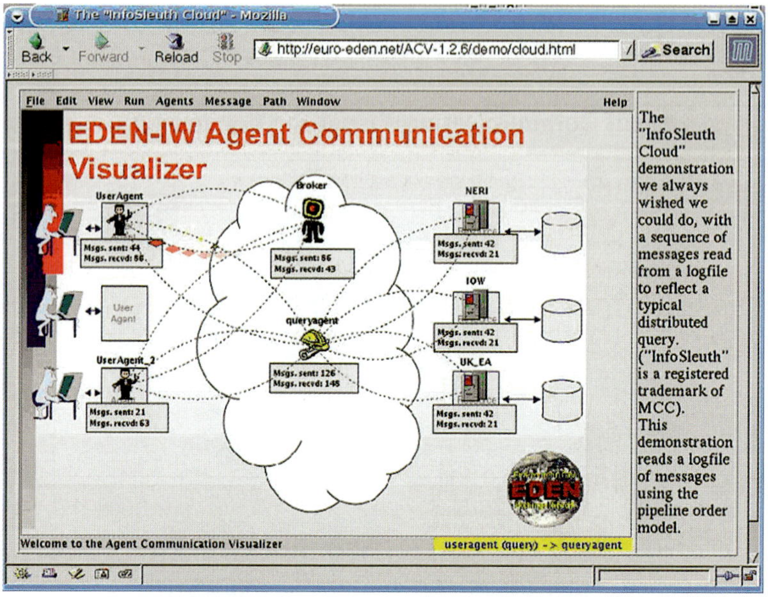

Fig. 11.3. The EDEN-IW agent system represented as a "cloud" of agents.

messaging is to be used for display. The "event.model" specifies the source of message and agent events. The poorly named "model.url" specifies source information for the event model. The "wait.interval.ms" specifies how long the ACV should allow for initial graphics layout before commencing message display, and "milliseconds.per.ms" indicates how long a millisecond of real time should be stretched in near real time.

order.model = simple or pipeline or rpc or jade
event.model = random or logfile or network
model.url = filename (for logfile event model)
model.url = port (for network event model)
wait.interval.ms = <numeric value>
milliseconds.per.ms = <numeric value>

The "order.model" property refers to the ordering semantics desired for display of messages. There are four conversation order models currently implemented: "Simple", "pipeline", "rpc", and "jade". The default is "simple".

order.model=simple
Each message is displayed in the order it is received by the ACV.

order.model=pipeline
Each message is displayed in the order it is received, but in addition, it is assumed that each message finishes before the next message starts.

order.model=rpc
Enforces the logical semantics of RPC. Details are available in the javadoc for the API of org.edeniw.dispatcher.RPCConversationHandler.

order.model=jade
Enforces the logical semantics of JADE as used in EDEN-IW. Further details are available in the javadoc for the API of org.edeniw.dispatcher.JadeConversationHandler.

Another "order.model" that was contemplated, but not implemented, is the *order.model = lamport*, so named for Leslie Lamport's axioms for event ordering in distributed systems. The Lamport model would provide logical ordering for systems in which messages that are monitored contain a sender ordering number and a receiver ordering number.

In the logfile event model, messages are read from a file, with the following fields delimited by tabs:

<counter><from agent><to agent><message type><message content>

This is the same format in which messages are saved to files. The log file from which to read the messages is specified by the property

model.url=logfilename

Note that although the name of the property is "url", the current implementation expects the filename to be a local file system's pathname.

In the network model, messages are read from a network socket with the same behaviour as logfile, delimited as one message to a line, with the same tab-delimited fields. The socket on which to listen for messages is specified by the property "model.url". Note that although the name of the property is "url", the current implementation expects simply a port number. The default port is 1997.

Because computer network message traffic often occurs within time spans shorter than can be visually comprehended by humans, the ability to display messages in a meaningful approximation to real time is important. For this reason, the applet also permits one to adjust the speed at which messages travel between agents, as well as how much time should be used to represent a millisecond of real time during the display of message traffic, in order to represent pauses between agent messages appropriately. To permit the ACV to run on slower machines, one can also alter the animation refresh rate and how much time to allow to initialise the system before beginning to display messages.

The "wait.interval.ms" property defines how long (in milliseconds) the ACV will wait before starting to display messages. Adjusting this value can be important because initialisation requires some time, and if message display begins too quickly, the menus may not be painted for some time after message activity begins.

wait.interval.ms = 3000

The "milliseconds.per.ms" property defines how much to stretch a millisecond of real time so as to make it possible for the ACV to display and the human to view the messages. The default is

milliseconds.per.ms = 1000

Sample ACV Properties File

The principal display properties of the ACV can be controlled via the menu items while the ACV is running, or through modifying the properties file used during the launch of the application. Additional parameters may be controlled through this file. Documentation regarding the setting of parameters is also available on the "Eden-Euro" website (http://euro-eden.net).

Fragments of the ACV properties file are shown below to illustrate the flexibility of configuration. Global properties include the background image, appearance of the message, agent, and statistics legends, and presence and arrangement of multiple frames.

Messages can be scaled and coloured individually, and their shapes can be selected from a small set of standard options. The appearance of the messages can be specified on a per-type basis. If appearance attributes are not specified, they inherit the global default values. Two message types, "aggregate" and "unknown", are required for any configuration. Other message types are determined by the message type list.

The appearance of paths between agents can be specified through global default values for either "shape", "segmentation", "visibility", "background colour", or"traversal speed". These properties can be further specified for paths between specific agent pairs. Individual paths can be further specified with multiple segments to obtain more complex message travel patterns.

Agent properties include "scale", "visibility", "colour", "location", "display of message statistics", and "display of messages". These properties can be specified as global defaults, by agent type, and on a per-agent basis.

Agent types are specified in a list. Agents themselves are also listed. Their types must be specified in per-agent properties, with unspecified agents receiving the "unknown" type.

11.4.2 User Interface Customisation

The experience of the development team (which admittedly has the bias of intellectual investment in the project) is that configuration of new applications is remarkably easy. Having identified a background image and suitable icons for agents (Figure 11.4), one can simply run the ACV to allow the agents to be populated automatically as unknown agents, and then, using the edit mode, quickly arrange the agent icons in appropriate locations, choose the agent types and assign appropriate icons, and then save the configuration file for subsequent use. Certain elements of edit mode are not particularly user-friendly,

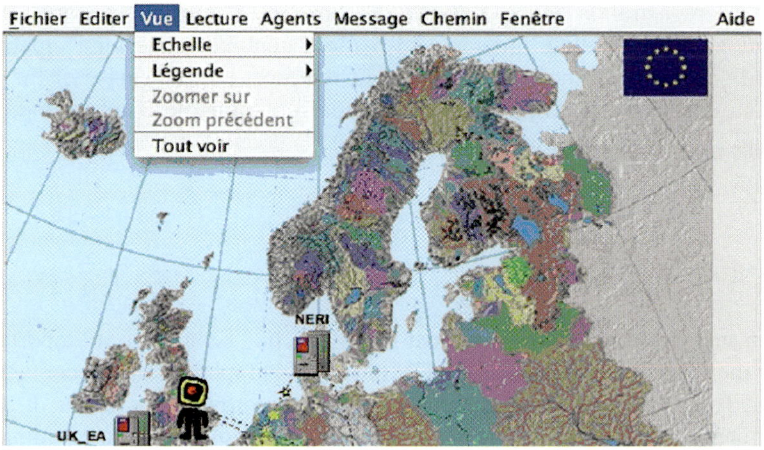

Fig. 11.4. A snapshot of the ACV showing French menus.

ergonomically speaking, but layout configuration is a fairly short activity, and those facile with text editors such as Emacs can edit the configuration file directly to reduce mouse mileage.

11.5 POTENTIAL USES OF THE ACV

The ACV was developed for the purpose of assisting developers in explaining the behaviour of agent systems. It has been demonstrated in that capacity and found useful.

In addition, the ACV has some use for debugging; however, a near–real-time message visualiser is only superficially useful for debugging distributed systems, since serious communication bugs often manifest themselves as timing-dependent problems, and precise timing is specifically outside the scope of the ACV. The use of the "VCR" replay mechanism may increase its utility for debugging, however.

Other possible uses include visualisation of process flow, as seen in the sequencing centre pipeline example on the ACV website, and display of social network activity.

11.6 THE ACV WEBSITE

The ACV software and source code were made publicly available under an open source licence during the final months of the EDEN-IW project at the following URL: http://euro-eden.net. The website provides several demonstration configurations of the applet, including a "space" demo showing some of the more sophisticated configuration possibilities;

the "cloud" demonstration and EDEN-IW demonstration using the Europe map; and a demonstration of the pipeline order model using a genome sequencing pipeline for its background. The website includes a zip file of the most recent release of the software, including an executable Java jar file, sample configuration files, source code, a user's manual, and application programmer interface (API) documentation.

11.7 CONCLUSIVE REMARKS

The design and development of an agent visualiser has been seen as important to the success of the EDEN-IW Project. The ACV provides a graphic means of conveying some of the underlying functionality to new users as well as providing a means of debugging the performance of the agent system. It is a valuable tool for demonstrations and for educating audiences on the unique type of functionality provided by agent systems.

The ACV prototype that has been developed also has extensibility beyond this project. The source code is to be made publicly available for others to use, adapt, and modify for visualising other agent-based systems. The ACV has been adapted to display the sequence of processes in a genome-sequencing pipeline. This code has been provided with the ACV as an available demonstration of alternative uses of the system and has been released with the ACV prototype source code.

```
####################
# Global message properties
#
message.scale = 1.0
message.shape = oval
message.color = 0x0

####################
# Per-type message properties
#
message.type.list = reply unknown aggregate query-ref
request agree inform ping

####################
# reply message properties
#
reply.message.shape = diamond
reply.message.color = 0xff3333
```

```
####################
# Global path properties
#
path.segments = single
path.shape = quadratic
path.when = always
path.display.type = single
path.traverse.ms = 2500
path.color = 0x404040
```

```
####################
# Per-type agent properties
#
agent.type.list = user directory task resource decision
unknown

##########
# task agent type properties
#
task.agent.image = worker.gif
task.agent.scale = 0.6
task.agent.visible = true
task.agent.label.location = west
task.agent.stats.location = south
task.agent.message.visible = true

# Individual agent properties
#
agent.list = task raioworacle raneri useragent_2
useragent_1 rawb raiow da rauk

task.name = task
task.external_id = task
task.type = task
task.x_coordinate = 284
task.y_coordinate = 499
task.agent.image = worker.gif
task.agent.scale = 0.6
task.agent.visible = true
```

```
task.agent.label.location = west
task.agent.stats.location = south
task.agent.message.visible = true
```

```
####################
# Global agent properties
#
agent.image = sleuthy.png
agent.scale = 1.0
agent.visible = true
agent.label.color = 0x0
agent.label.visible = true
agent.label.location = north
agent.stats.visible = true
agent.stats.location = south
agent.message.visible = true
```

```
##########################################
#
# Visualizer properties
#
####################
# Global properties
#
legend.foreground.color = 0x0
legend.background.color = 0xe1e1f0
legend.title.foreground.color = 0xffffff
legend.title.background.color = 0x0
stats.foreground.color = 0x0
stats.background.color = 0xe1e1f0
background.image.file = europe-map-01.jpg
animation.interval.ms = 100

####################
# Frame properties
#
frame.list = frame0
```

```
##########
# frame0 frame properties
#
frame0.frame.size.width = 656
frame0.frame.size.height = 674
frame0.frame.location.x = 0
frame0.frame.location.y = 22
frame0.frame.view.x = 0
frame0.frame.view.y = 0
frame0.frame.scale = 1.0717734625362931
frame0.frame.editing = true

####################
# Background properties
#
background.size.height = 600
background.size.width = 800
```

Chapter 12

INTERACTION WITH THE USER

A. Papageorgiou and J. Würtz

12.1 INTRODUCTION

This chapter, which describes the User Interface of the EDEN-IW prototype, is intended to both technical and non-technical readers. The former will be mostly interested in the first part of the chapter, which describes the most important design decisions, while the latter will be interested in the rest of the chapter, in which the capabilities of the EDEN-IW Web user interface are demonstrated, to give a feeling on the variety of possibilities the prototype offers to the user.

12.2 WEB APPLICATION DESIGN

The EDEN-IW Web application[1] was designed as a three-tier application (Figure 12.1). The presentation uses the Model-View-Controller (MVC) design pattern, which is implemented using the Jakarta Struts framework.[2]

The business layer holds all business logic. It consists of the following modules and agents:

- the analysis functions module, which is responsible for the data analysis functionality
- the cache module
- the user management module
- the User Agents
- other added-value services modules.

All business logic can be accessed through a set of business delegates.

Finally, the data layer is the local Web database. Please note that for simplification reasons the remote JADE Agent platform, which is responsible for getting the Inland Water data, can also be considered as part of the data layer. This simplification allows making the Agent system functionality and Inland Water databases access transparent to the Web application.

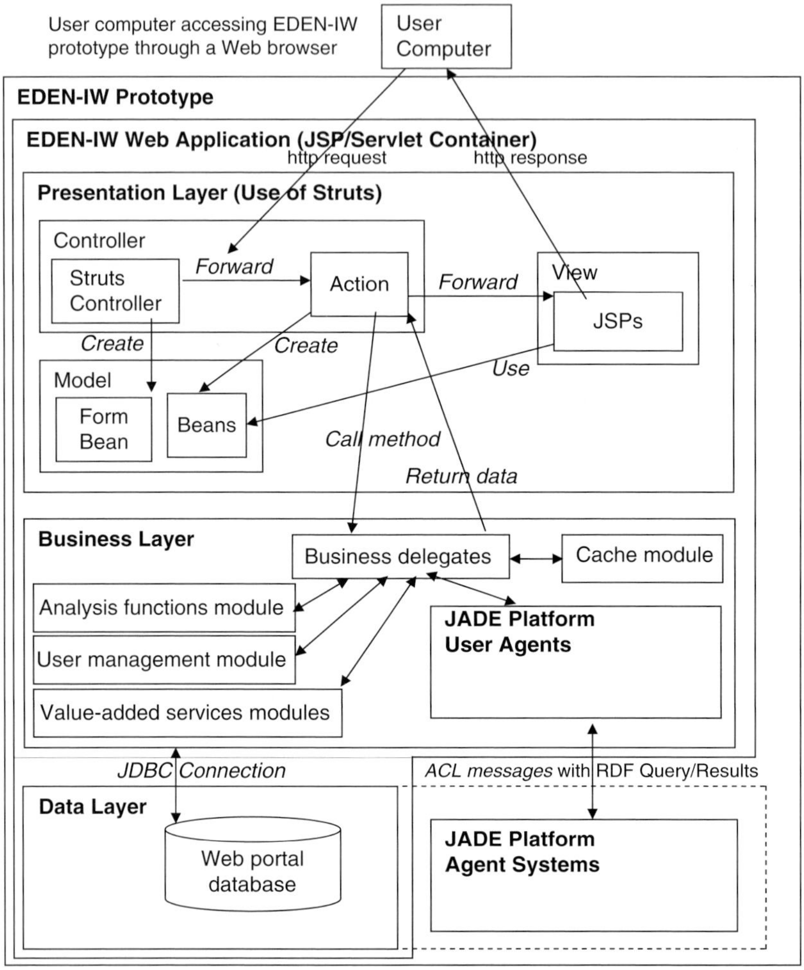

Fig. 12.1. The three layers in the EDEN-IW Web application.

12.2.1 Communication Between the Web Application and Agent System

All communication with the Agent System is realised through the User Agents using RDF messages that comply to the EDEN-IW ontology. There are two important aspects in the communication between the User Agent and the Web application:

- The first one is the methodology to be used for making method calls and exchanging data. JADE agents are allowed to communicate with non-agent classes either by invoking them themselves (i.e. the actor is the agent) or by receiving objects from them (i.e. the agent is the receiver of a request). Therefore, in order for the Web application classes to communicate with the User Agents, a shared object was used.

- The second one is the behaviour that should be used for handling the communication. JADE agents allow either a blocking or non-blocking behaviour. This means that either the application should wait for the User Agent to respond (blocking) or it will continue as soon as it sends a request (non-blocking).

The first behaviour is non-acceptable as in EDEN-IW many requests could take longer than a few seconds, which would mean that the user waits without receiving any feedback. The second behaviour is acceptable as it quickly returns feedback to the user that his/her request is pending; however, in requests that are answered in much less than a second, this feedback is useless as when the users sees it, the request is already answered and he/she has to refresh the browser to see the results, thus adding an unnecessary step to the process.

To overcome the drawback of the non-blocking behaviour, a new "block until timeout" behaviour was implemented. By setting the time-out to the time that the user can wait for a respond before he/she receives feedback, the problem is solved.

The entire process of communication between User Agent (UA) and Web application (Figure 12.2) is as follows:

1. The UA[3] listens for objects which have been sent to it (shared DTOs, i.e. data transfer objects).
2. The application user makes a request to the Web application.
3. The Web application creates shared Query/Request DTOs (one or more) that hold all information necessary for the UA.
4. The Web application sends one shared object (DTO) to the UA, waits to see if the request is handled within the predefined time-out and, if not,[4] it continues showing an empty page to the user (with a message like: "please wait while your request is being processed by the system") that is automatically refreshed every *X* seconds.
5. The UA makes, upon receiving a request, the necessary tasks to process the request. After the User Agent receives the query, it transforms it into RDF format, and passes it to the Task Agent. The interaction[5] between the Task Agent and the Resource Agent(s) will provide the former with the data satisfying the query to pass back to the User Agent. The data itself is not processed in the User Agent, but from the Web application.
6. During this time the page is refreshed. Additional information is only presented if the UA updates the DTO with status information.
7. When the UA finished, it updates the shared DTO.
8. In the next refresh of the Web page, the Web application will see the new data in the shared DTO and will update the HTML output.

12.2.2 Analysis Functions Module

Having solved the retrieval of Inland Water data from heterogeneous distributed databases, the second step is the analysis of this data. The data analysis functions were designed in order to allow easy expansion of the functions without any need to change the Web

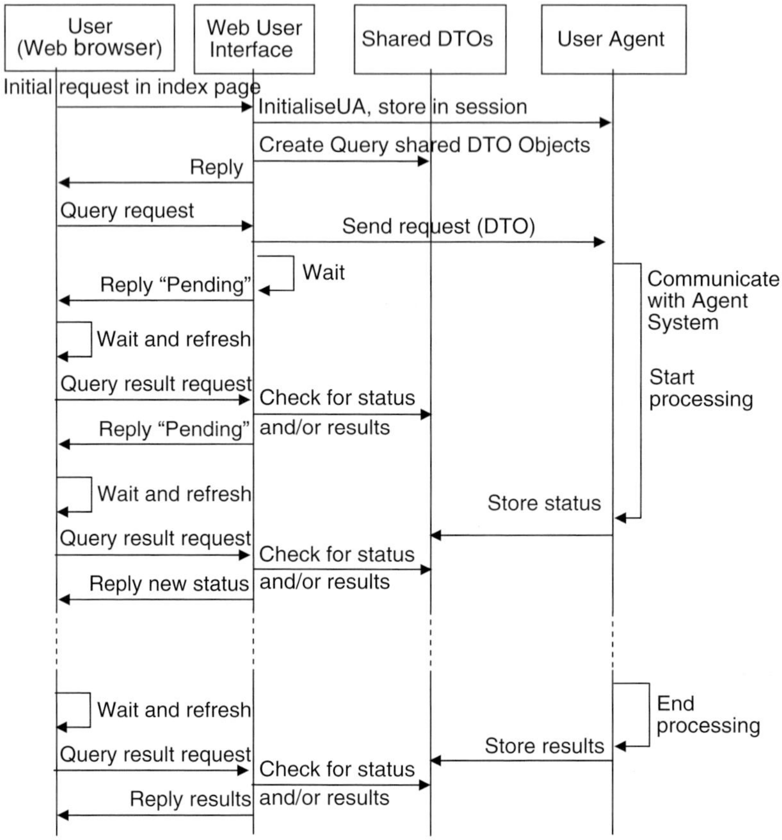

Fig. 12.2. Communication process between Web interface and user agent.

application. This is achieved through the definition of the analysis function registry and the analysis function interface. This is illustrated in Figure 12.3:

1. The registry reads the available analysis functions from a property file, and initiates them.
2. When the Web application analysis is requested on a set of data, the registry asks each available analysis function if it can process the data.
3. Each function replies if it can process the data, a description on the analysis function, and any additional input data it requires.
4. The user is shown a list of available functions that can process the data, as well as a list of the functions that cannot process the data.
5. When the user has selected an analysis function, the data are forwarded to the related function. Here they are processed, and the results are then returned back and presented to the user.

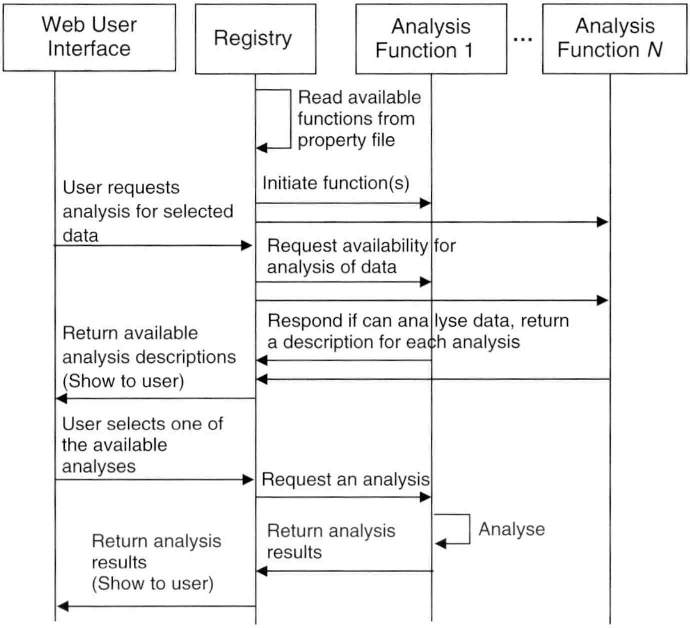

Fig. 12.3. Communication process between Web interface and analysis function modules.

12.2.3 Cache Module

Due to the fact that (a) the data are distributed, (b) a single query can request a large amount of data form multiple sources, and (c) certain queries could take considerable time, it was deemed as necessary to implement a data cache.

The cache is implemented twofold:

1. A caching mechanism is implemented in the server memory. This mechanism is using a LRU (Least Recently Used) algorithm for deleting old data.
2. As an additional local database that uses a schema that has a one-to-one match with the EDEN-IW ontology. Deleting old data is based on the date of their last update and/or an LRU algorithm in cases where the database size reaches a predefined threshold.[6]

In both cases every user query, prior to being sent to the User Agent, is checked against the cache. If the results are found there, then the User Agent is not used.

Caching significantly improved the performance of the application, because it reduced the Internet traffic between the User Agent and the rest of the agent system. Based on measurements,[7] the average mean time for retrieving data for a particular station, determinant and analytical fraction from around 2 seconds was reduced to less than 100 milliseconds for cached data.

12.2.4 User Management and Value-Added Services

The implementation of the basic EDEN-IW functionality (data retrieval and data analysis) was further enhanced with value-added services that enhanced the EDEN-IW concept. Additional modules could easily be added as new components to the existing application.

User management

Under the user management component, all user-related functions are grouped, such as authentication, user editing, and changing of password.

Scheduler

Certain functionalities require processes to be executed at particular intervals (e.g. off-line download and cache management). The scheduler component is responsible for invoking the processed whenever necessary based on time-schedule data stored in a property file.

Mailer

The mailer component is used to send e-mails to the registered EDEN-IW users, who use the off-line download or notification services.

Ontology Service

The Ontology Service is responsible for parsing all the ontology DAML and OWL files. It also creates Java Objects, and supports methods that give access to the ontology classes.

Glossary

The Glossary component connects to the database in order to retrieve glossary information that should be presented to the user.

12.3 WEB APPLICATION OVERVIEW

12.3.1 The Search Interface

The Search Interface (Figure 12.4) contains the query elements. Based on these, the users have various choices to formulate their queries. The choices include:

- The "station selection", which can be accomplished through:
  - the EDEN-IW GIS map interface pop-up window
  - a search based on determinants pop-up window
  - simple text input pop-up window
- All the above-mentioned methods populate the station list of the Web page, from which the user can select stations to perform his/her query.

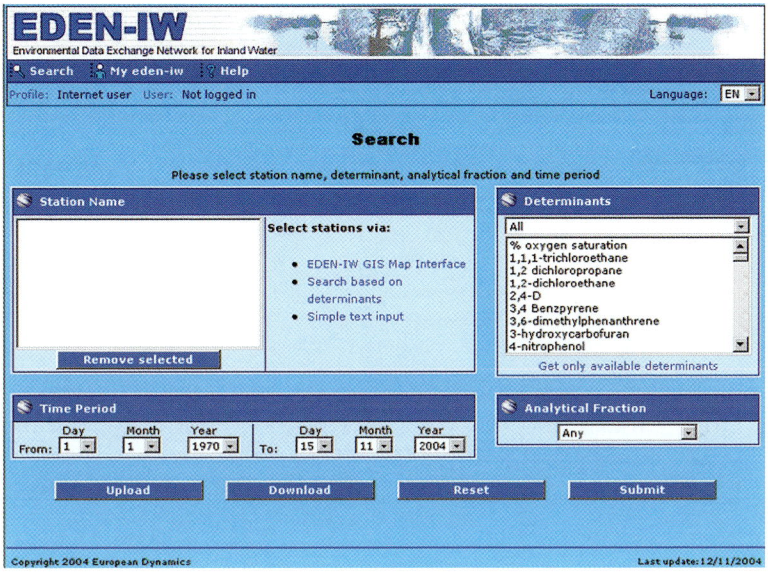

Fig. 12.4. The user's selections are displayed on the interface; here in English.

- The "determinant selection", which can be accomplished by selecting a determinant group (top drop-down selection list), and the related determinants from the bottom list. The lists initially contain all the determinants available in the EDEN-IW global ontology. By pressing the link "Get only available determinants", the list decreases to the determinants that are contained in the Directory Agent.
- The "time period selection", which allows the user to select the period of the EDEN-IW query.
- The "analytical fraction selection", which allows the user to select the analytical fraction of the EDEN-IW query.

All the above selections are mandatory for the correct formulation of the EDEN-IW query. In case some data is absent or invalid, an alert is activated for the user to try again.

12.3.2 Multilingual User Interface

The EDEN-IW prototype features a multilingual interface. Currently, three languages are supported:

1. English (Figure 12.4)
2. French (Figure 12.5)
3. Danish (Figure 12.6).

The language is automatically determined from the user's browser language settings, but it can also be changed from application. The system can be easily extended to accommodate more languages by modifying a single configuration file. From the three

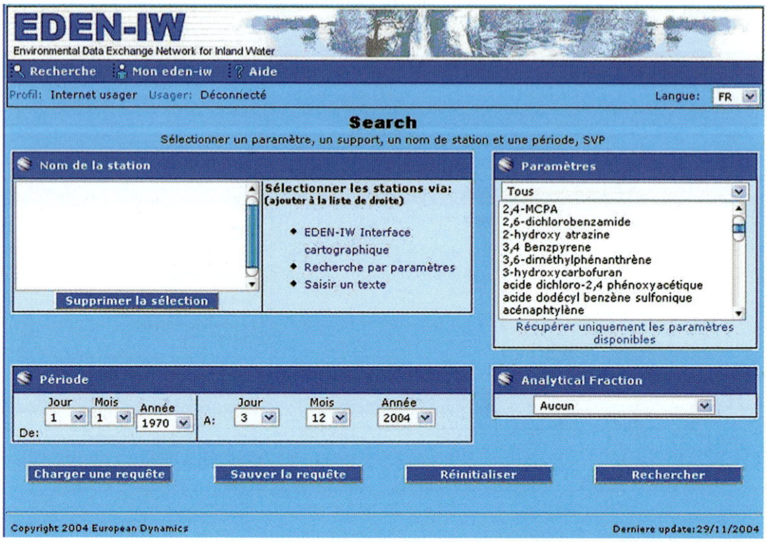

Fig. 12.5. Detail from French user interface.

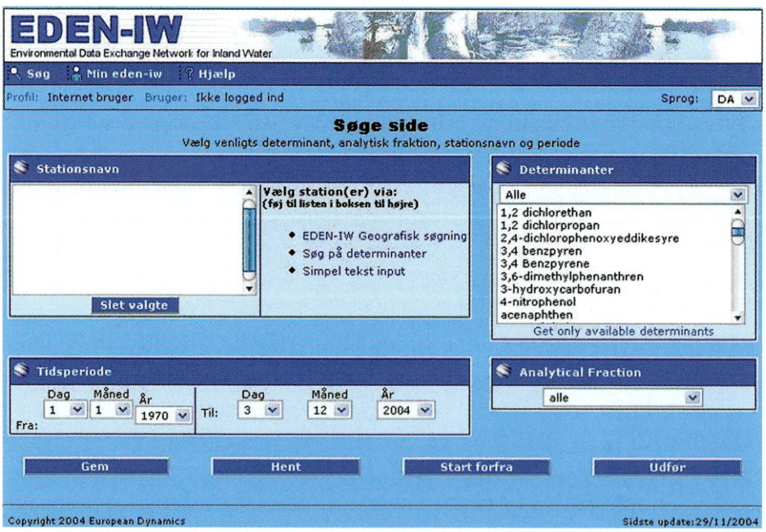

Fig. 12.6. Detail from Danish user interface.

screenshots it can be observed that the various textual data are correctly translated in the appropriate language.

12.3.3 The Results Interface

The query, formed by the user's selections, is submitted and received at this stage by the User Agent. The results screen is refreshed as long as there are still pending data. When the User Agent returns all the requested data, the user interface displays the results of the queries, in graph form or in standard tabular form (Figures 12.7 and 12.8).

12.3.4 Analysis Functions

The results can be analysed further by any of the functions described in Table 12.1. The analysis results can also be added in the list of search results, and thus they can be displayed in graphical format or be retrofitted in an analysis function.

12.3.5 User Management and Value-Added Services

User management is not a critical aspect of EDEN-IW application. However, it is an important issue for future developments, as it will allow the support of personalised and value-added services. Table 12.2 summarises the EDEN-IW profiles available and related functionality.

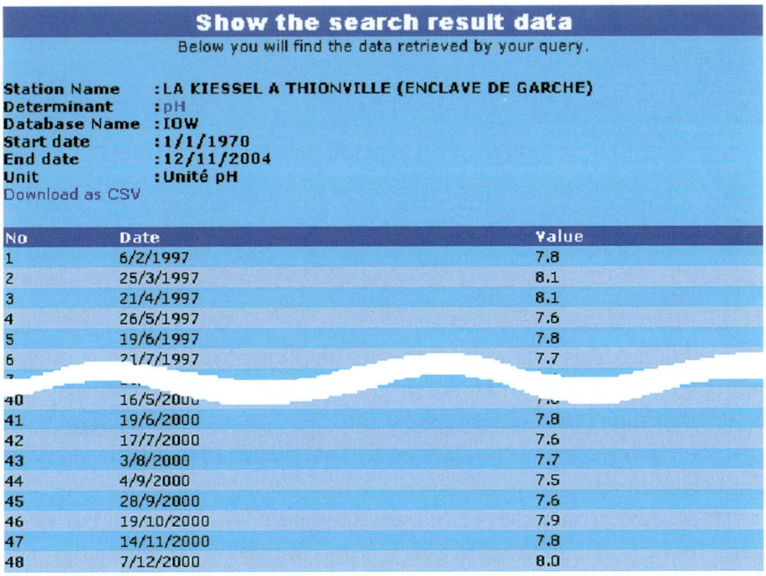

No	Date	Value
1	6/2/1997	7.8
2	25/3/1997	8.1
3	21/4/1997	8.1
4	26/5/1997	7.6
5	19/6/1997	7.8
6	21/7/1997	7.7
[illegible]	[illegible]	[illegible]
40	16/5/200	[illegible]
41	19/6/2000	7.8
42	17/7/2000	7.6
43	3/8/2000	7.7
44	4/9/2000	7.5
45	28/9/2000	7.6
46	19/10/2000	7.9
47	14/11/2000	7.8
48	7/12/2000	8.0

Fig. 12.7. Numerical results.

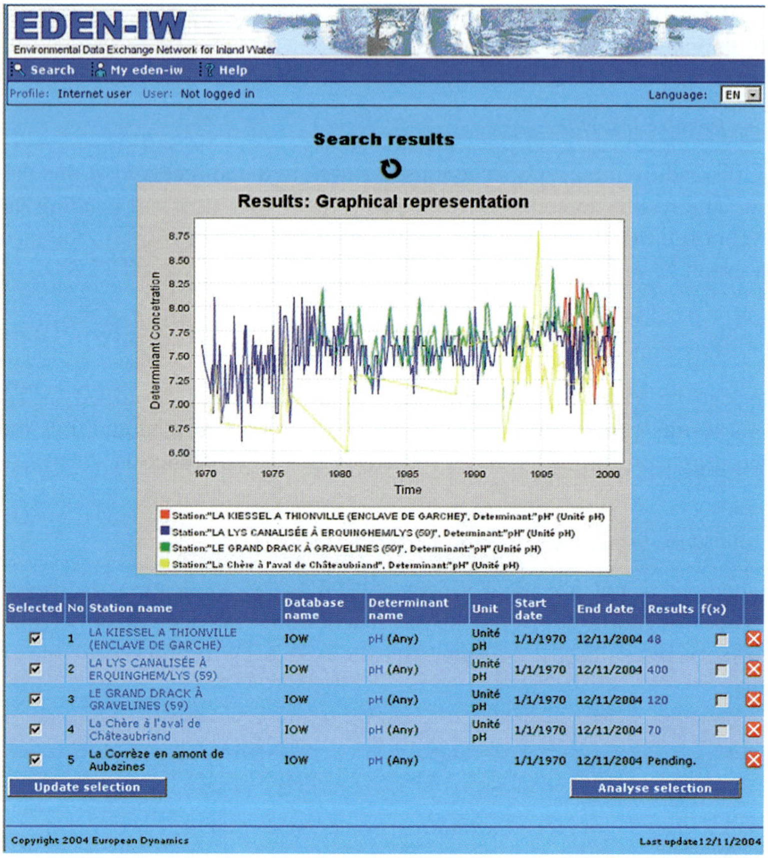

Fig. 12.8. Graphical display.

The value-added services include:

- *User management*: A user, based on his/her role, can update, insert, or delete user data.
- *Off-line download*: A user can request the off-line download of a large number of data. An e-mail is sent to the user when the process finishes. Such downloads are scheduled during hours that the application usage is low.
- *Saved queries*: Each logged user's queries are saved and are available every time he/she uses the application again.
- *Notification*: A user can select to be interested on particular queries. The system checks periodically for new data in the query results and notifies accordingly the user via mail.
- *Download/Upload*: Downloading a query locally in XML format, and uploading a locally saved query is available to all users.

- *Glossary*: The EDEN-IW glossary is integrated into the Web application and is available to all users.
- *Help*: A help page is available to all users.

Table 12.1. EDEN-IW analysis functions

Number of Time History Curves	Analyse Function	Options	Output
>0 with same unit	Unit Conversion	mg/l < − > kg/l, etc.	Time Histories Converted
>0 with same determinant and unit	Statistical Analysis		Mean Value and Standard Deviation as Constant Time Histories
>0 with same determinant and unit	Determinant Conversion	Nitrate < − > Total Nitrogen, etc.	Time Histories Converted
1	Time Projection	1 month–5 years after last measurement date	Linear Time History
1	Measurements Transformation	Log(x), ln(x), exp(x), 10^x, x + const, x − const, x*const, x/const	Time History Transformed
1	Shift Measurements Dates	Up to +/ − 40 years	Time History Shifted
2	Correlation		Correlation Coefficient as Constant Time History
2 with same determinants and units	Mathematical Functions	Addition, Subtraction (1 − 2, 2 − 1) Multiplication, Division (1/2, 2/1)	Time History Transformed
2	Mathematical Functions	Multiplication, Division (1/2, 2/1)	Time History Transformed
>2	Mathematical Functions	Multiplication	Time History Transformed
>2 with same determinants and units	Mathematical Functions	Addition, Multiplication	Time History Transformed

Table 12.2. EDEN-IW profile functionalities

Profile	User management	No. of active queries	Off-line download	Saved queries	Notifi-cations
Internet User Any user accessing EDEN-IW via the Web	No	10	No	No	No
EDEN-IW User An authenticated EDEN-IW User, with minimum value-added functionalities	Self	30	No	Yes	No
EDEN-IW Power User An authenticated EDEN-IW User, with full value-added functionalities	Self	100	Yes	Yes	Yes
EDEN-IW Administrator An authenticated EDEN-IW User, with full value-added functionalities and no query limit (necessary only for testing purposes). This user can create/delete/edit other user accounts	All	Any	Yes	Yes	Yes

12.4 USING EDEN-IW FOR INLAND WATER ANALYSIS

In the following sections, the most important analysis functionalities are illustrated. The functionalities should be regarded as a first step towards the implementation of a decision support system as outlined in Würtz (2003).

12.4.1 Selecting the Initial Inland Water Data

Figure 12.9 shows the search page of EDEN-IW prototype, where the search for stations is based on a specific determinant. In this case we are looking for stations with the determinant “pH”. We also require that the measured value (at a certain time in the time series) is above 7.

The result of the search is a series of stations. We select some of these, and return to the Start page of the prototype. We then select the time period and press the “Submit” button. This is shown in the Figure 12.10 below.

The result of the search is shown in Figure 12.11. Here the Station Name, Database Name, Determinant Name, Unit, Start Date, and End Date, as well as the number of measurements are given.

The time series are automatically plotted while the data are searched. Afterwards, the user can deselect if desirable, and then pressing the “Update Selection” button.

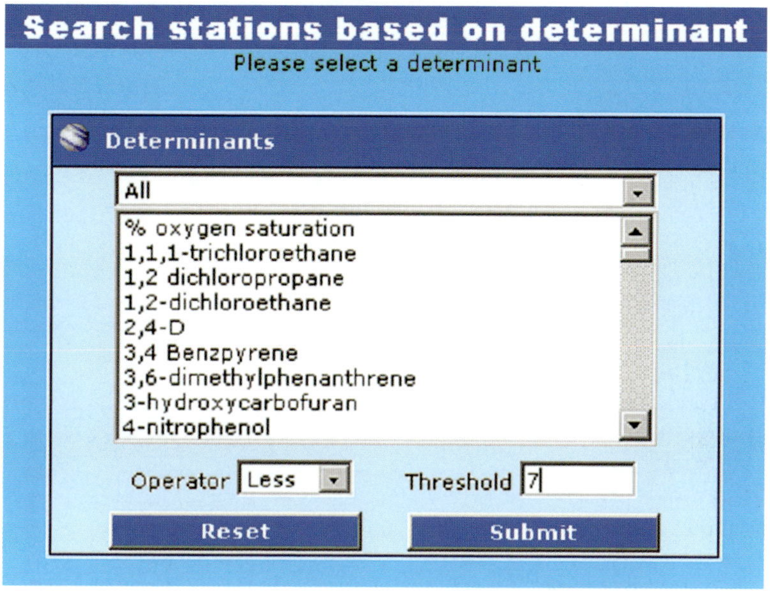

Fig. 12.9. Search for stations with a specific determinant.

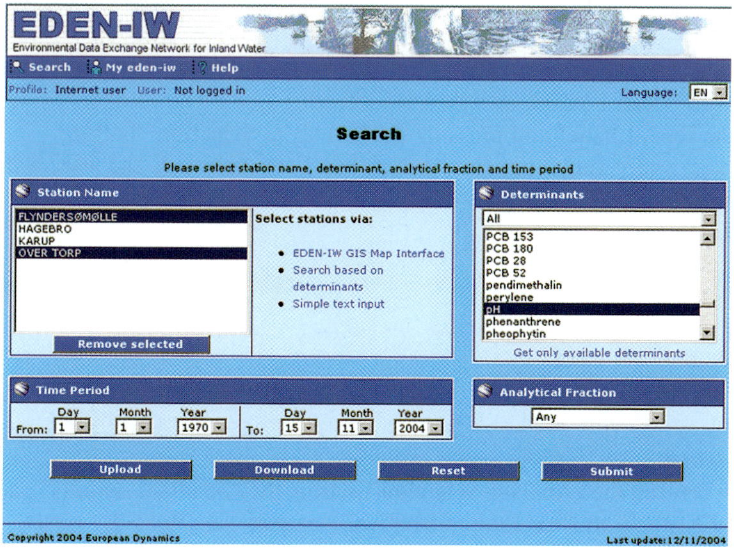

Fig. 12.10. Search for pH data.

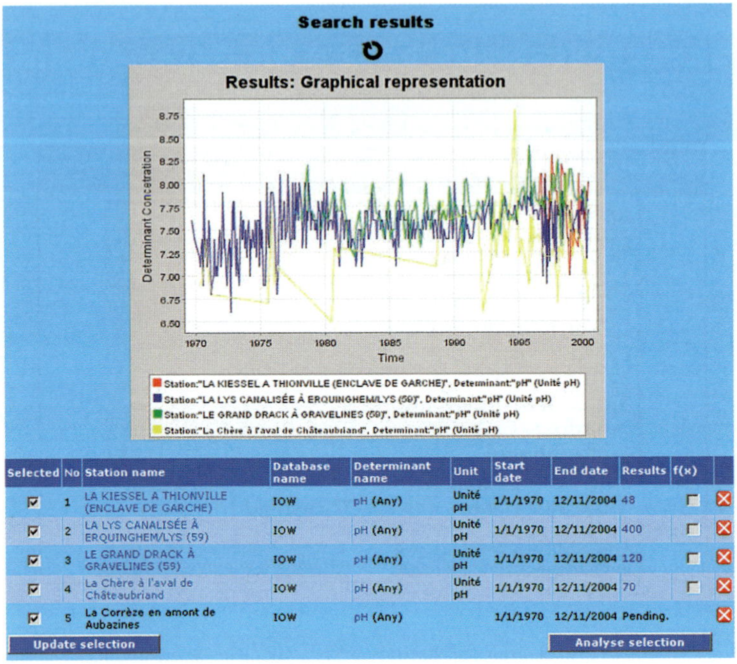

Fig. 12.11. Presentation of measurement results.

12.4.2 Time Series Analysis for One Time Series

These time series can now be analysed by selecting one or more of them in the "$f(x)$" column in the Result Panel and pressing the "Analyse Selection" button.

In the case where only one time series has been selected, the user will have the following choices (see Table 12.1):

- Statistical values
- Time projection
- Determinant conversion
- Measurement transformation
- Shift measurements dates
- Unit conversion.

Statistical values

After having pressed the "Analyse Selection" button, the "Analysis options" panel appears (see Figure 12.12). In this case the "Statistical Analysis" is selected.

The result of the analysis is shown in Figure 12.13. Here the mean value and standard deviation are given. These values are also given as constant time series with only two data points representing the (constant) value and the start and end dates. The standard

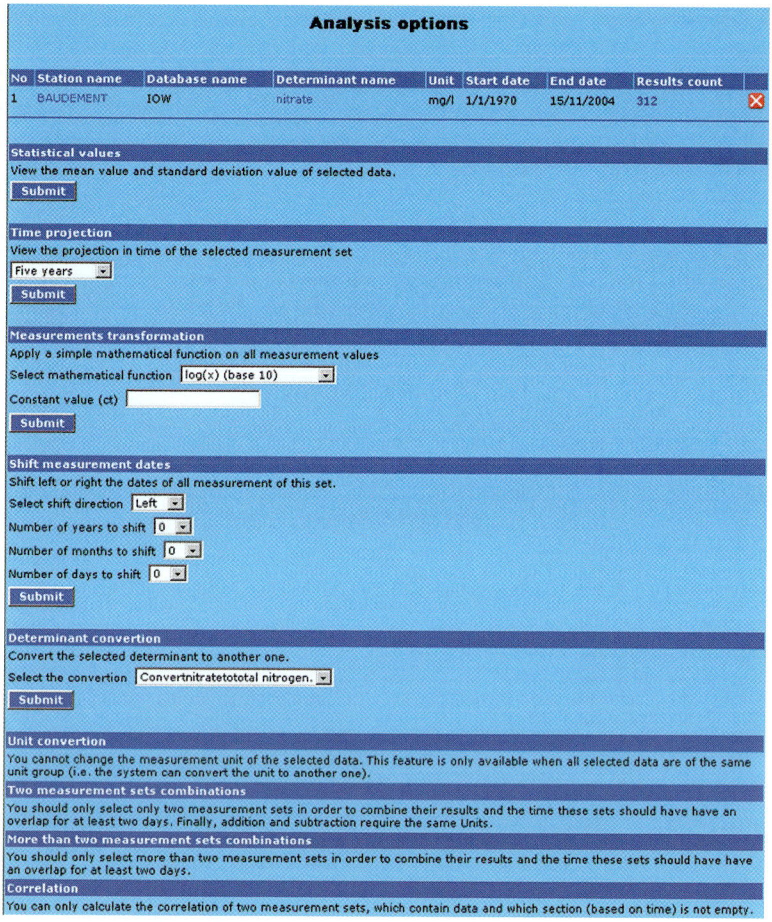

Fig. 12.12. Selection of statistical analysis.

deviation is given by the 2 constant value time series equal to the mean value +/− the standard deviation. In the figure the mean value is selected to be saved. This is then selected in the previous "Presentation of Measurement Results"panel, and then plotted together with the time series. This is shown in Figure 12.14.

Projecting in time

The time projection is done by means of a linear regression analysis of the data. The user can select the projection time (calculated from the last data point date) to be in the range from one month to five years. In Figure 12.15 the influence of the analysis period for the time series is illustrated. Here the second projection takes into account only the most recent data points.

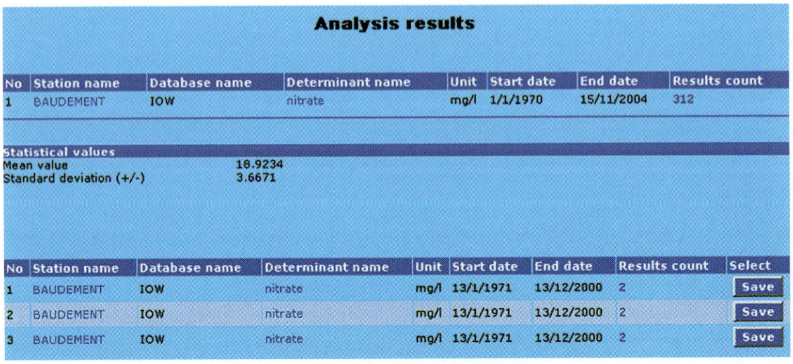

Fig. 12.13. Analysis results.

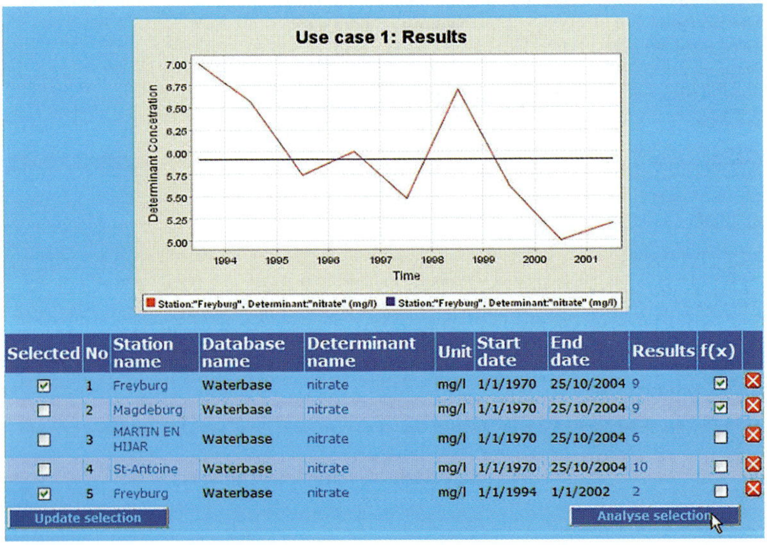

Fig. 12.14. Plotting the mean value.

Determinant conversion

In the case of the selection of some specific determinants, it is possible to make a conversion of the measurement values to another determinant. This is, for example, the case with Nitrate, which can be converted to Total Nitrogen (or vice versa) by assuming that all the Nitrogen is in the form of Nitrate. By applying such a conversion, the possibilities of comparing with relevant data from other databases increase. In Figure 12.16 this conversion is selected in the "Analysis Options" panel (in fact for three time series with

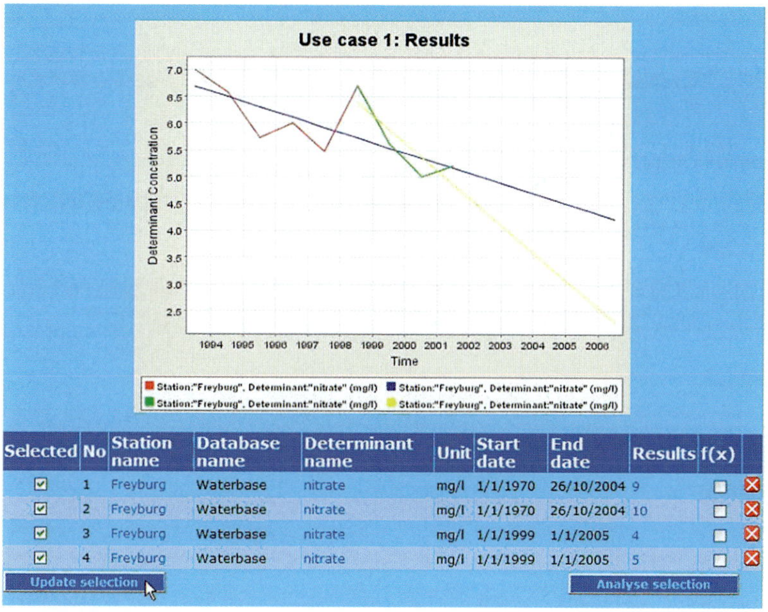

Fig. 12.15. Projection in time. The influence of the analysis period on the projection result is illustrated.

the same determinant and unit), and the result is plotted together with the original data in Figure 12.17.

Mathematical operations
It is possible to make simple mathematical operations on the single time series. This option can be very useful when constructing indicators from the original data. The user has a series of mathematical operations at his/her disposition (see Table 12.1).

12.4.3 Time Series Analysis for Two Time Series

In case two time series are selected for analysis, the user has the following possibilities depending on the Determinant and/or the Unit are the same or not in the two series:

- Statistical analysis (if same Determinant and Unit)
- Determinant conversion (if same Determinant and Unit)
- Unit conversion (if same Unit)
- Correlation (if a time overlap exists)
- Mathematical operations: Addition and Subtraction (if same Determinant and Units and a time overlap exists)
- Mathematical operations: Multiplication and Division (if a time overlap exists).

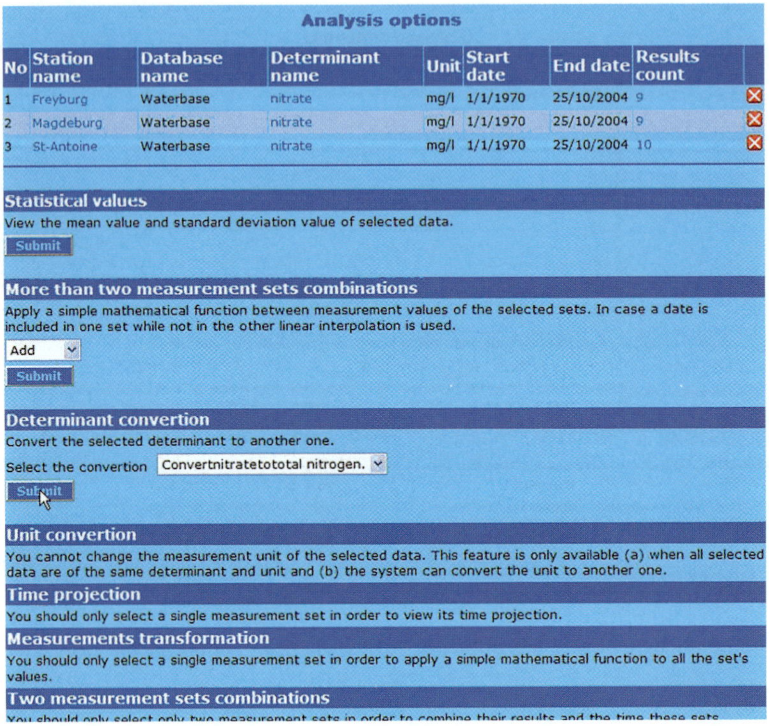

Fig. 12.16. Selection of determinant conversion from Nitrate to Total Nitrogen.

Correlation

The value of the correlation coefficient varies between 0 and 1, where 1 indicates that the two series are fully (linear) correlated.

The correlation could e.g. be applied at two stations on the same river to find the position of a larger source of pollution. If the time series are highly correlated, the source is most probably to be found upstream both stations.

The correlation could also be applied to two time series at the same station but with different determinants, to investigate if these determinants are originating from the same source of pollution.

In Figure 12.18 is shown the time series of two stations at the same river (Elbe) for the same determinant Nitrate. The analysis show that the time series are highly correlated with a coefficient $r = 0.947$.

12.4.4 Calculation of Load/Pressure Indicators

To be able to answer the question "Where to clean first?" (i.e. "Clean where most beneficial"), a relevant indicator could be "Load/Pressure". Here "Load" could be the

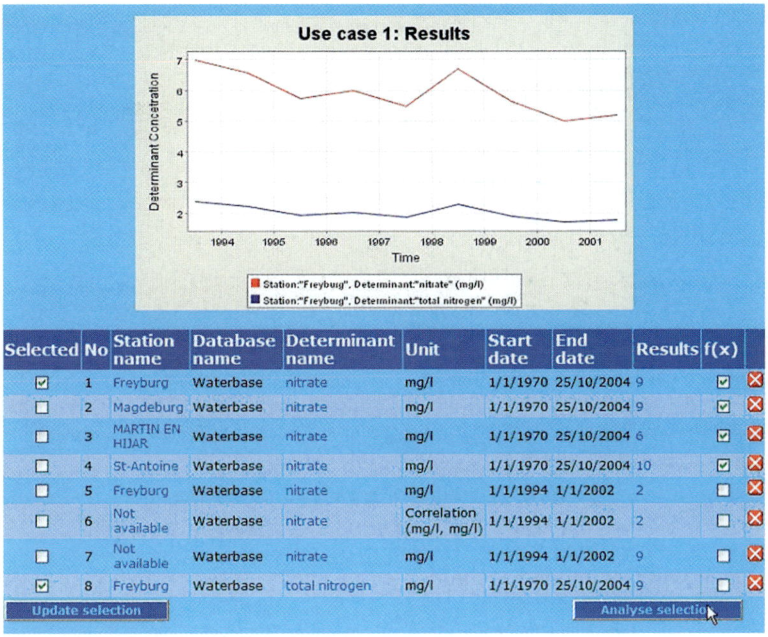

Fig. 12.17. Presentation of the converted results (Total Nitrogen) together with the original data (Nitrate).

Table 12.3. Determinands for the load/pressure indicator

Determinand	Unit	Table
Average total N	mg/l N	Quality_Rivers
Discharge	m^3/s	Quantity_Water or Stations_Rivers
Catchment area	km^2	Quantity_Stations
Population density	capita/km^2	Pressures_Rivers

Average Annual Concentration of e.g. Nitrogen multiplied by the Annual Discharge, and "Pressure" could be the Population upstream in the catchment.

The higher the value of the indicator, the more urgent it is to clean the river upstream.

In order to calculate this indicator we need the following determinants in Table 12.3 at a specific station and aggregation period (e.g. yearly) in the WaterBase:

12.5 FINAL REMARKS

It is seen that the project has developed a series of possible user interactions with the databases and the data. These are presented to the user in a standard browser, and the

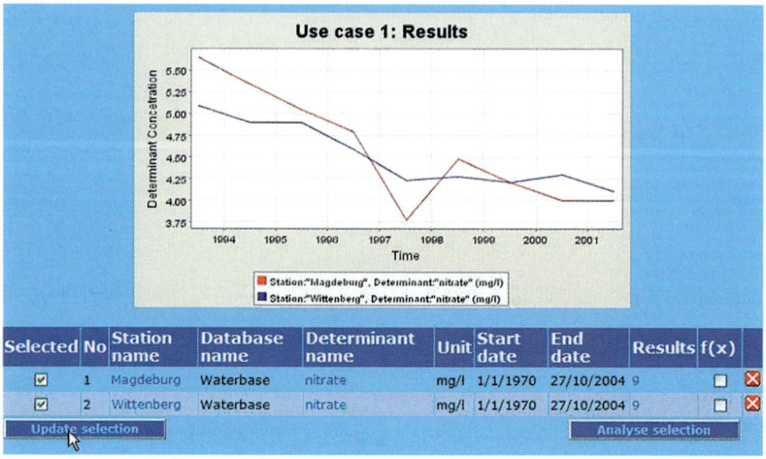

Fig. 12.18. Two highly correlated time series.

use of the underlying combined technologies of agent software and Web tool are thus transparent to the user.

NOTES

[1] A Servlet/JSP container is used for running the application. Any container that supports the Servlet 2.3 and JSP 1.2 specifications will suffice (such as Tomcat 4.1.31).

[2] Apache Jakarta Struts: http://struts.apache.org/

[3] Each EDEN-IW user has his/her own user agent.

[4] If yes, this means that step 7 is completed and thus the application executes step 8.

[5] See Chapter 8. describing the Agent System for more details.

[6] The threshold depends largely on the underlying infrastructure.

[7] Measurements results largely depend largely on the topology of the EDEN-IW system installation and the underlying infrastructure. A set of PCs (Pentium 4, 3 GHz, 512 RAM) connected via Ethernet was used for the measurement.

Chapter 13

GEOGRAPHICAL INFORMATION AND MAP SERVICES

F.-X. Prunayre

13.1 INTRODUCTION

This chapter describes the technology behind the map service implementation in the EDEN-IW user interface.

The amount of geographical data dealt with by the users is continuously increasing. As a consequence, access to Geographical Information Systems (GIS) in Web applications becomes more and more important.

This is the case, both for the user input, for instance the choice of a given measurement station, and for the user output, where result can be linked to a given regional area or river basin.

As will be demonstrated in this chapter, a GIS system connected to a map server offers a very user-friendly way to access and choose the geographical input needed in the EDEN-IW application, and increases the user friendliness and functionality of the application.

13.2 GIS DATA IN THE INLAND-WATER DOMAIN

13.2.1 Data Types

The water domain is using a lot of geographic information in vector format (river network, monitoring stations, catchments, etc.) or raster format (Digital Terrain Model [DEM], etc.).

The EDEN-IW project is using monitoring stations as basis for the analysis as shown in the user interface. Primary GIS requirements for the EDEN-IW application are to display the main features – at least stations from different databases over Europe – and to be able to make basic GIS searches.

To realize those two main requirements, a lot of emerging technologies could be used. These technologies are based on IT standards mainly coming from the Open Geospatial Consortium, OGC (2005), and the technical committee for Geographic information/Geomatics of ISO (ISO/TC 211, 2005).

13.2.2 Common Ways of Distribution

Geographic data over the Internet is available in many different ways. Web mapping interfaces have been developed by all well-known commercial GIS software, such as ESRI, MapInfo, IONIC, and GeoConcept during the last 12 years. Also emerging open-source applications have been developed to build such interfaces. The main limitation of those Web-mapping applications was the inability to distribute information in any format other than an image in an HTML page.

The distribution process takes place in a four-step process:

1. *Search*: requires a huge effort (time and knowledge of data provider) to be able to find what fits the needs
2. *Collect* (download, order, etc.): requires finding out the explanation of what are the terms of use of the datasets
3. *Convert*: requires data processing and harmonization
4. *Use*.

Due to these recurrent problems in gathering geographic information, standards have been developed on methods, tools, and services for data management, including definition and description. Metadata conforming to ISO-19115 help data description, catalogues help data discovery, and geoservices help in accessing data from the level where they are maintained most effectively.

13.3 OPEN GEOSPATIAL CONSORTIUM STANDARDS

Two main Web services have been specified by the OGC to allow geographic data to be accessed remotely in read or read and write access modes. The Web Map Service (WMS) specification provides four protocols (GetCapabilities, GetMap, GetFeatureInfo, and DescribeLayer) to create maps. These maps may be provided by multiple distributed and heterogeneous sources. The Web Feature Service (WFS) specification describes how to manipulate features (following the Simple feature specification of the OGC) in order to establish communication between client and server at the level of the entity.

The query needs to define the following:

- Coordinate system (SRS)
- Bounding box
- Output format
- Filter (geographic or attribute).

The coordinate system is most of the time described using European Petroleum Survey Group code (OGP, 2005). The bounding box is mentioned in the WGS84 coordinate system. The output format is qualified using MIME/TYPE and the filter in XML.

The Web Map Service returns images in different formats (PNG, JPEG, etc.) and is based on HTTP. Using WMS, a client is able to combine images coming from multiple distributed and heterogeneous servers (Figure 13.1).

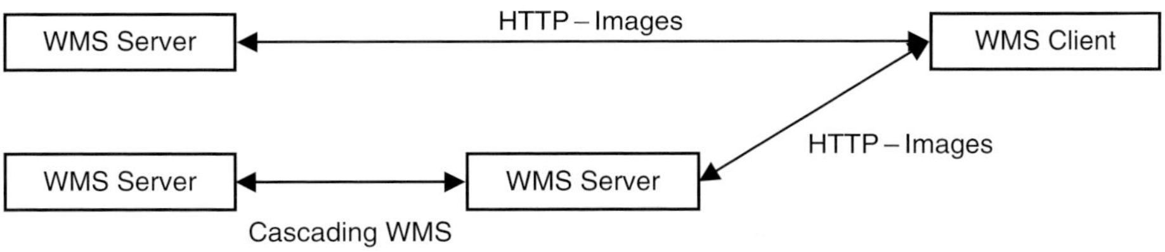

Fig. 13.1. WMS services.

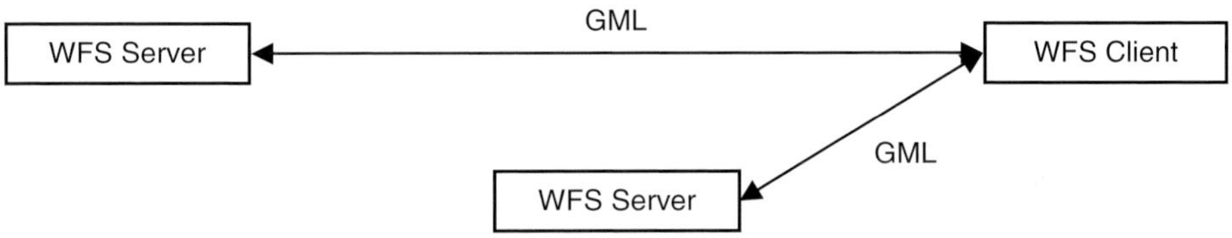

Fig. 13.2. WFS Services.

Queries specified in the WMS context are

- GetCapabilities (XML service description)
- GetMap (image)
- GetFeatureInfo (optional).

The Web Feature Service returns vector objects and uses GML for encoding of data (Figure 13.2).

Queries specified in the WFS context are

- Basic WFS (i.e. Read access)
  - GetCapabilities
  - DescribeFeatureType
  - GetFeature
- Transactional WFS (i.e. Read and Write access)
  - Transaction (create, update, delete)
  - LockFeature.

13.4 EDEN-IW GIS SYSTEM ARCHITECTURE

The EDEN-IW Agent System has already been described in detail in Chapter 8, and in Chapter 12 the functionality of the User Interface was detailed. Here we focus on the additional Web services implemented for the map server.

13.4.1 Concepts

No datasets are owned by the main platform. The main platform acts as a portal for distributed services over Europe. The communication between nodes uses OGC standards (Figure 13.3):

- WFS for databases connected to the EDEN-IW system
- WMS for datasets used as background maps.

Data transfers are all based on HTTP using REST (or SOAP).

For WMS, the WMS server converts data from local format (ESRI shapefile, PostGIS, Oracle Spatial, Mapinfo TAB, etc.) to an image according to the coordinate system, bounding box, and output format described in the query. For WFS, the WFS server converts data from local format to GML according to the coordinate system, bounding box, and output format described in the query.

13.4.2 Geographic Data Perspective

The possible themes may be grouped into two parts: the positions of the stations in the connected databases, and several orientation layers.

National borders are always present for orientation. As a basis of interoperability, the system is able to plug all orientation layers served by the OGC WMS server.

Subsets of the common orientation layer are already registered into the system. The user can distinguish between general orientation layers having a pan-European extent and local datasets, which should be adapted according to the position of the map during navigation (Figure 13.4). EDEN-IW databases layers (served by WFS) are the following:

- Monitoring stations for Denmark
- Monitoring stations for UK

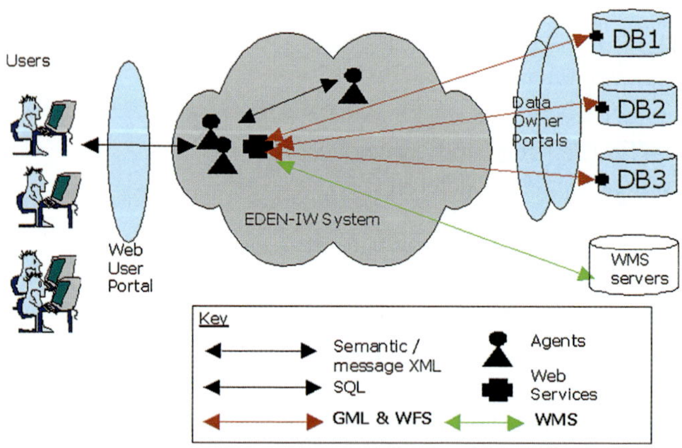

Fig. 13.3. WMS and WFS communication in EDEN-IW.

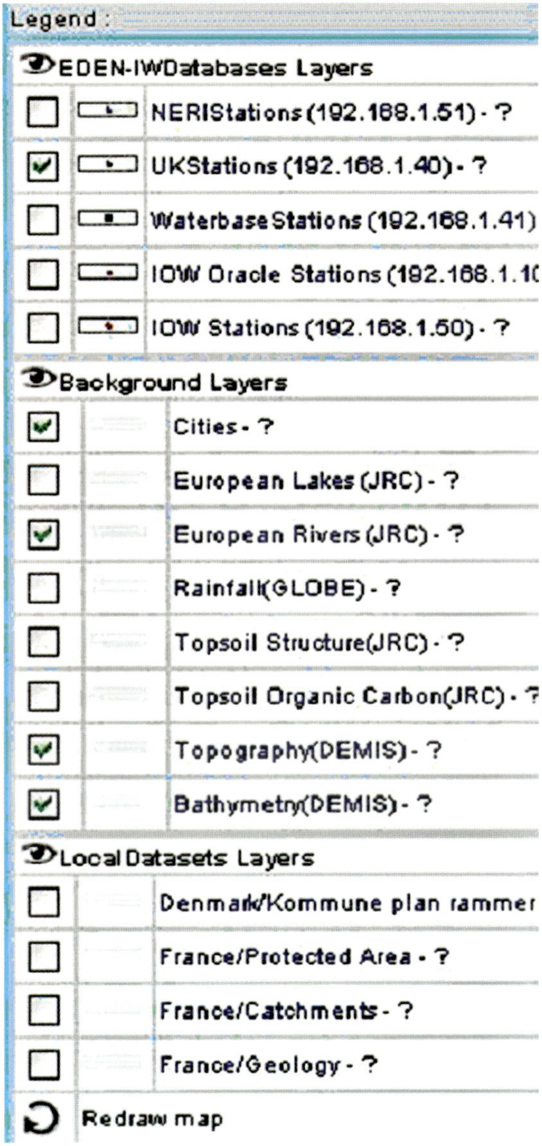

Fig. 13.4. EDEN-IW database and orientation layers.

- Monitoring stations for Waterbase
- Monitoring stations for France (Full dataset)
- Monitoring stations for France (subset).

Orientation layers (served by WMS) for Europe are the following:

- Cities
- European lakes from the JRC
- European rivers from the JRC
- Rainfall from the GOLBE project
- Topsoil structure from the JRC
- Topsoil Organic Carbon from the JRC
- Topography from DEMIS
- Bathymetry from DEMIS.

Orientation layers (served by WMS) for large scale view are the following:

- Administrative area for Denmark
- Protected area for France
- Catchments for France
- Geology for France.

In the legend panel, a hyperlink to the metadata for the orientation layer is also found (if provided by the data owner). Metadata and catalogue processing are less relevant for the GIS part of the EDEN-IW prototype, but metadata integration in SDI is the main entry point for all usage. Recent catalogue specification will allow harmonizing the processing of distributed metadata.

13.4.3 User Perspective

The GIS search interface opens with a view of Europe (Figures 13.5 and 13.6). In the right panel is a table of contents that allows the user to select the themes to be presented in the map view. The upper panel contains the GIS tools for zooming in and out and moving to the centre of view. This panel also has shortcuts to named regions in a drop-down list. Below the map view is the clicking area for the selection of the stations contained in the current view.

The GIS user interface provides the following features (Figures 13.7, 13.8, and 13.9):

- Zooming and panning
- Zoom to predefined places such as country or catchments over Europe
- Layer manager allowing to manage layer visibility and to access metadata
- Overview map
- Search tool to select stations close to a point on the map and in a rectangle.

The selection of the station(s) can be made with the use of the map graphical interface whereby the user can zoom in/out on a particular region of Europe (all European countries and main catchments are currently featured) and select a particular station or group of stations in the current window. In Figure 13.7 are shown stations from selected databases and in Figure 13.9 is shown the selection of a specific catchment to be displayed.

In Figure 13.10, an example of the selection of all displayed stations is shown, together with the overview map. An example of a list of the selected stations is shown

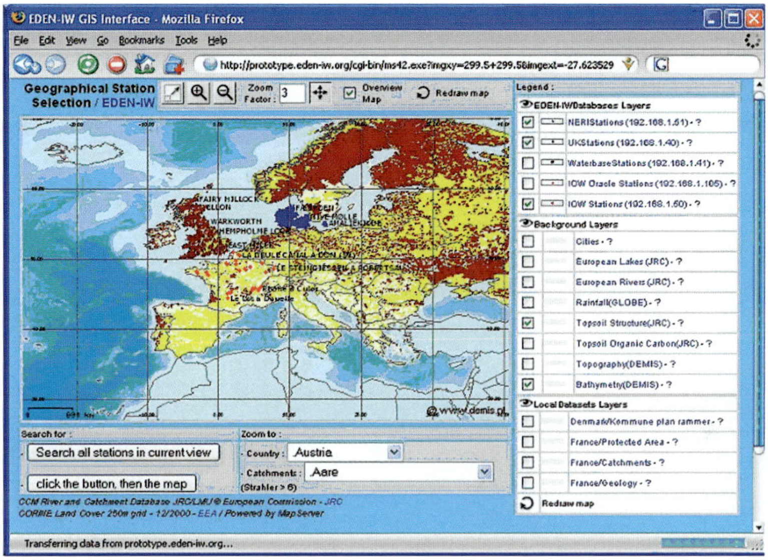

Fig. 13.5. User interface with pan-European orientation layers.

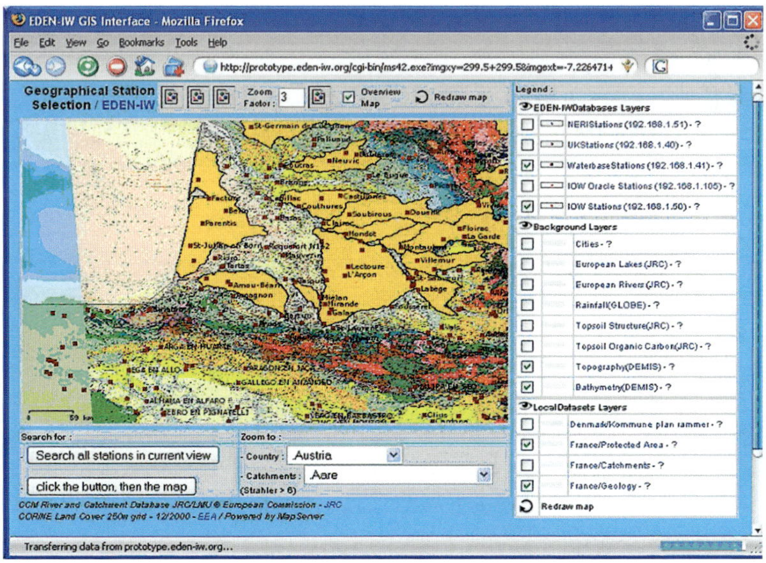

Fig. 13.6. User interface with local French orientation layers.

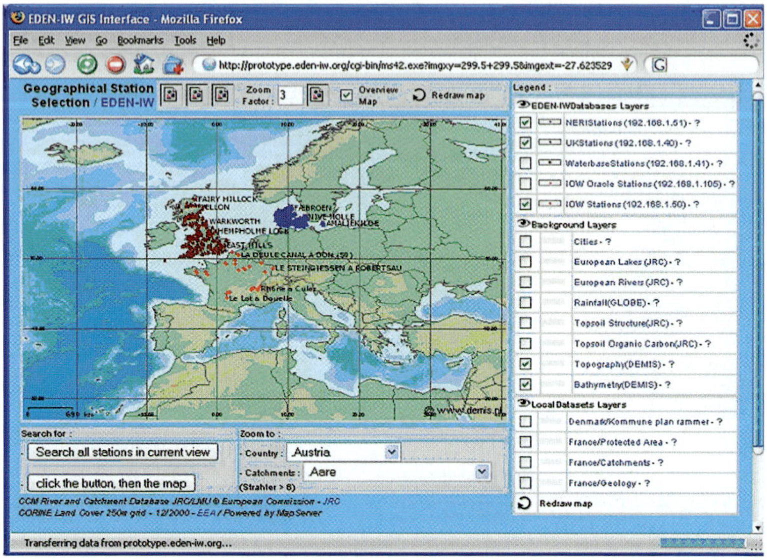

Fig. 13.7. Display of stations from selected database layers.

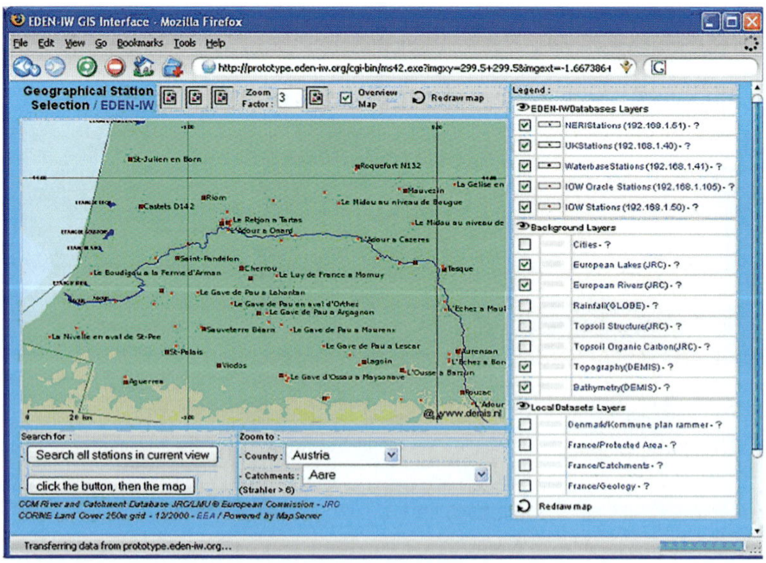

Fig. 13.8. Result of a zooming action.

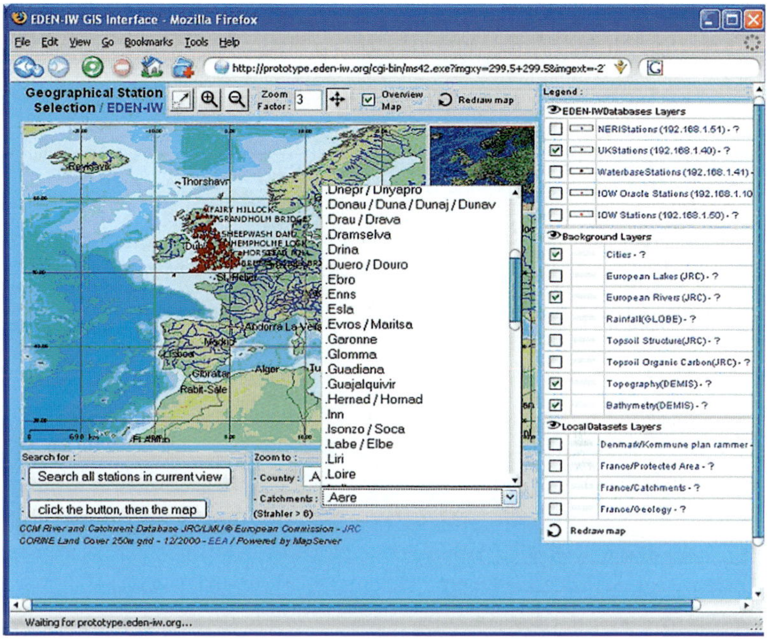

Fig. 13.9. Selection of a catchment.

in Figure 13.11. Having selected the station(s) of interest from the list, the user can then select the other parameters necessary for the search in the main panel of the EDEN-IW interface.

13.4.4 Software Perspective

The software used on the EDEN-IW database nodes originated from an open-source project called MapServer (2005). This project is a development environment for constructing spatially enabled Internet Web applications. MapServer runs on each node of the prototype, as detailed in Table 13.1.

Parts of the dataset are accessible through the user interface coming from a remote GIS data server. In this case, the system relies only on the OGC standard used to access the datasets and not on software used to publish them. However, some incompatibility remains due to the missing finalization of the specification.

All datasets used for the background map come from different GIS data servers using WMS and different software (Mapserver, ESRI ArcIMS, IONIC, Mapxtrem, etc.; Table 13.2).

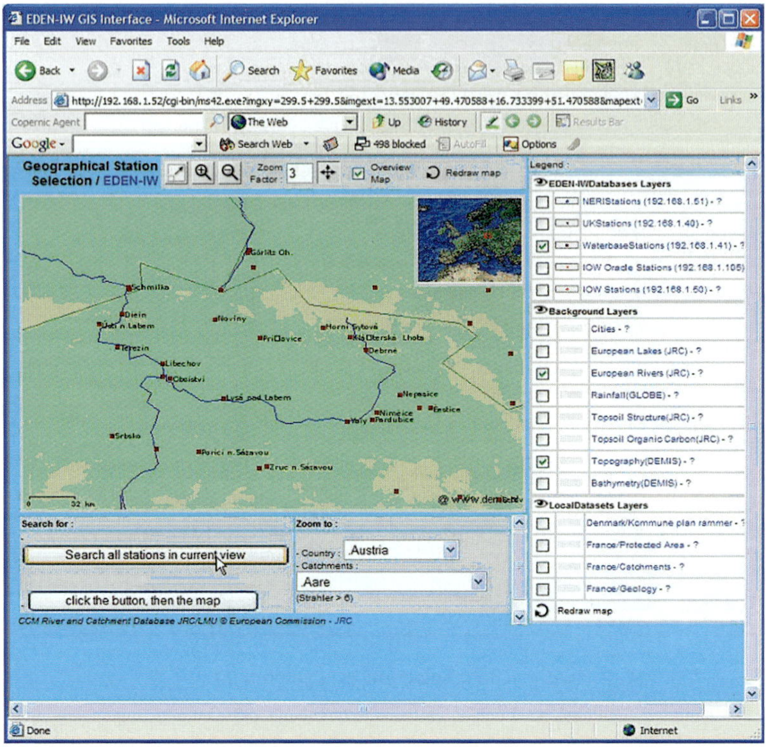

Fig. 13.10. Selection of all displayed stations.

13.4.5 *Performance*

Architectures based on interoperability are (de facto) distributed architectures, which rely on network availability and flows. The main problem observed by the EDEN-IW prototype is the transfer of vector data using GML and WFS. The version of WFS used shows incompatibility issues depending on the software used. The latest WFS specification with the GML 3–Level 0 profile should help the exchange of data between users. With the monitoring stations, which are point data, the file could be bigger than 1 MB – depending on the map requested. In such situations, it could be relevant to implement data compression between the client and servers to limit the data transfer size.

Security issues should also be taken into account, depending on the data the system deals with. However, services in the GIS fields are all based on REST or SOAP protocols, which are using HTTP and port 80 for communication. This port is commonly open for firewalls.

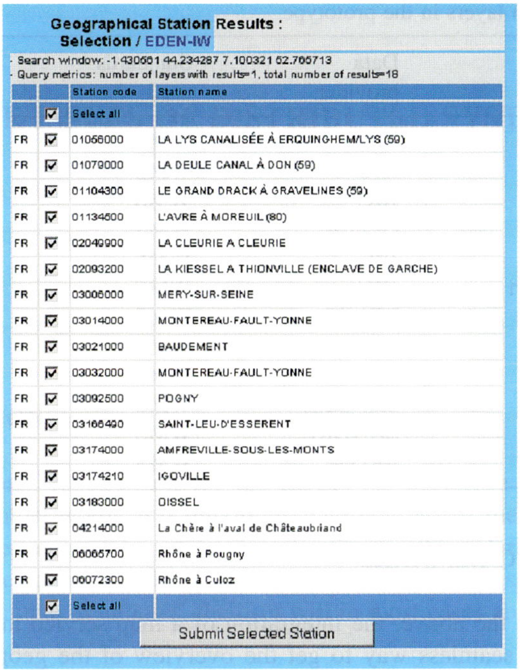

Fig. 13.11. Result of a station search.

Table 13.1. MapServer applications in the prototype

	GIS functions	Standards used	Data format read	Data format published
User Interface	Interaction with user Access remote server	WMS & WFS client	GML & WMS images	Images
IOW access node	Serve monitoring stations through HTTP	WFS server	Microsoft Access database using ODBC	GML
IOW Oracle node	Serve monitoring stations through HTTP	WFS server	Oracle Spatial using OGR*	GML
NERI node	Serve monitoring stations through HTTP	WFS server	Microsoft Access database using ODBC	GML
UK-HMS node	Serve monitoring stations through HTTP	WFS server	Microsoft Access database using ODBC	GML
Waterbase node	Serve monitoring stations through HTTP	WFS server	Microsoft Access database using ODBC	GML

* OGR (2005) is a library used by MapServer (and other OpenSource projects) to read different vector formats. OGR is a component of GDAL (2005).

Table 14.1. Main database characteristics

Database	No. of stations	No. of measurement records
NERI	534	346 380
IOW	30	92 278
Waterbase	3 438	1 89 253
UK-HMS	277	5 65 225

Initially, a wireless 11 Mbps hub was used, but due to technical problems this approach was abandoned. It was found that sometimes the wireless connection caused the agents to hang or even exit during a query. It is possible that these difficulties were due to noise or non-expected input in the message contents. In many cases, it was necessary to restart the agent platform(s), to remediate the problems. Sometimes it was also necessary to restart the main agent platform to reset the prototype. As similar problems were experienced when the agent platforms were connected via the Internet, this issue definitely has to be addressed more thoroughly in the future.

The cause of the difficulties was not resolved, but the situation improved with a wired and thus higher speed connection.

However, it should be stressed that the EDEN-IW Java agent platforms are based on a version of JADE (version 2.5), which also by other developers were found to have difficulties interacting with the http protocol. An improved version (version 3.3) of the JADE framework is now available, where these problems should have been solved.

It should finally be mentioned that in the wired intranet configuration, the system was stable for several days without requiring any restarts.

14.3 ANALYSIS METHOD

As mentioned above, a Linux PC was connected to the network to monitor the traffic. Here the UNIX executable *tcpdump* was applied with the options:

```
tcpdump –A –n –s 0 > dumpfile
```

where
A prints out the packet in ASCII (when possible)
n keeps the hostnames as IP numbers (they are not converted to names via DNS)
s 0 sets the minimum packet length to 0 so all packets are captured in full length.

A sample fragment of the outputfile (*dumpfile*) is shown in Table 14.2.

The file shows the start of a message from IP 192.168.1.253:4643 (the platform with RAUK on channel 4643) to IP 192.168.1.231:3043 (the TA on channel 3043 on the Main Platform). It covers in total 3.5 milliseconds from the time 16:18:52.768349 to 16:18:52.771858. After some initial packets (and the last part of a message in the other

Table 14.2. Fragment of an output file from the tcpdump command

```
16:18:52.768455 IP 192.168.1.253.4643 > 192.168.1.231.3043: P 1:8(7) ack 1 win 17520
E../q<@....X.........#....=.....P.Dph}..JRMI..K

16:18:52.769140 IP 192.168.1.231.3043 > 192.168.1.253.4643: P 1:21(20) ack 8 win 17513
E..<9@...<}...........#......=.P.Di....N.192.168.1.253...#

16:18:52.769368 IP 192.168.1.253.4643 > 192.168.1.231.3043: P 8:27(19) ack 21 win 17500
E..;q=@....K.........#....=.....P.D\.....192.168.1.253..

16:18:52.769946 IP 192.168.1.253.4643 > 192.168.1.231.3043: P 27:534(507) ack 21 win
17500
E..#q>@....b.........#....=.....P.D\.y..P....w"................@..............F.sr..jade.lang.acl.
ACL Message6..........I..performativeJ..reply_byInMillisec[..byteSequenceContentt..[BL..
contentt..Ljava/lang/StringBuffer;L..conversation_idq.~..L..destst..Ljade/util/leap/ArrayList;
L..encodingq. ~..L..in_reply_toq.~..L..languageq.~..L..messageEnvelopet.*Ljade/domain/
FIPAAgentManagement/Envelope;L..ontologyq.~..L..protocolq.~..L..reply_to q.~..L.reply_
withq.~..L..sourcet..Ljade/core/AID;L..userDefPropst..Ljade/util/leap/
Properties;pxp............
```

direction at time 16:18:52.769140), the ACL message content in XML starts in packets of 1024 bytes.

By sorting the content of the output files after IP addresses, one can get an overview of the timing of the messages. In our case, the sorting was based on the receiving IP address, so a global string search for, for example, ">192.168.1.253." generated a file with all the packets sent to the RAUK. The timing can then be estimated by importing the time stamps of the initial and final packets of a given message from this file into an Excel spreadsheet (Table 14.3).

The transmission time is then calculated as the difference between the time stamps of the final and the initial packets. The processing time needed by a specific agent or service to generate a new message from a received one is then calculated as the difference between the time stamps of the initial packet of the generated message and the final packet of the received message.

An example is shown in Table 14.3 from a Use Case 9 Request, "Which stations have data on pH?" In this case all the four databases have data, so all the four resource agents reply. Notice that the processing time of the four RAs is in parallel, so the times do not add up, but the total time is dominated by the longest of the four processing times.

Because all the agent platforms are connected in a LAN, the transmission speed is more or less constant (around 1 MBps), and rather small in comparison to the processing times. In the "real world", with the platforms connected via the Internet, the situation could be very different.

Finally, it should be mentioned that the overhead to the system due to the monitoring PC was found to be negligible. Repetition of the same query with and without the Linux PC connected to the network showed an increase of the total elapsed time of only 1–2%.

Table 14.3. Example of process and transmission times calculation

Time (hh:mm:ss.ssssss)	From->To	No. of bytes P:Process/T:Transmission	Time (h:mm:ss.sss)
16:18:55.997248	UI->UA		
16:18:56.215905		P: UA + TA	0:00:00.022
16:18:56.237517	TA->RANERI		
16:18:56.287748		13280 T:TA->RANERI	0:00:00.050
16:18:56.238352	TA->RAUK		
16:18:56.260753		13280 T : TA->RAUK	0:00:00.023
16:18:56.242430	TA->RAWB		
16:18:56.327146		13276 T : TA->RAWB	0:00:00.085
16:18:56.258284	TA->RAIOW		
16:18:56.274199		13278 T : TA->RAIOW	0:00:00.016
		P:RAWB	0:00:58.826
		P:RANERI	0:00:27.772
		P:RAIOW	0:00:01.166
		P:RAUK	0:00:18.072
16:18:57.439884	RAIOW->TA		
16:18:57.488303		46740 T : RAIOW->TA	0:00:00.048
16:19:14.332825	RAUK->TA		
16:19:14.863515		410039 T : RAUK->TA	0:00:00.531
16:19:24.059525	RANERI->TA		
16:19:24.818783		749229 T : RANERI->TA	0:00:00.759
16:19:55.153025	RAWB->TA		
16:19:59.105614		3821189 T : RAWB->TA	0:00:03.953
		P : TA + UA	0:00:12.122
16:20:11.228115	UA->UI		
16:20:11.668152		T : UA->UI	0:00:00.440
		TOTAL	0:01:15.671

14.4 THE SEARCH FOR AVAILABLE DETERMINANDS

The first search concerns the search for available determinands. This is initiated when the user clicks on "Get only available Determinands" on the home page of the User Interface (see Figure 14.1).

When a resource agent registers at the directory agent, it informs about available determinands in the related database.

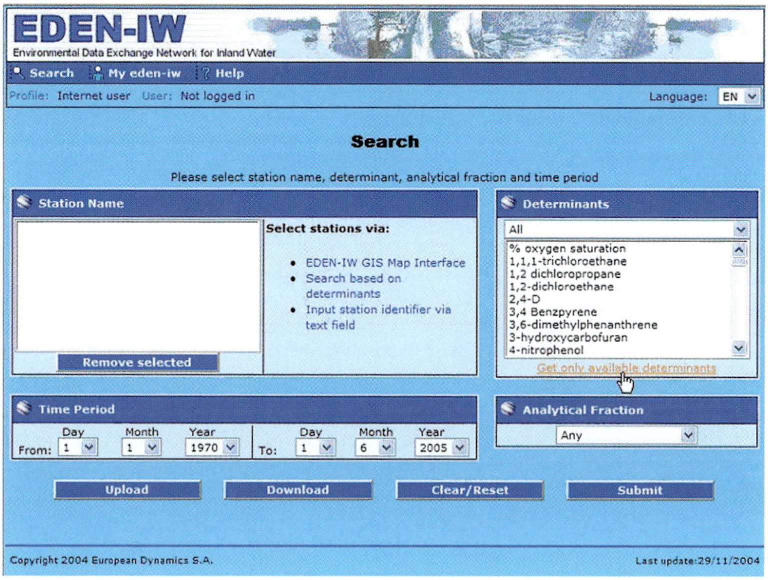

Fig. 14.1. The home page of the user interface.

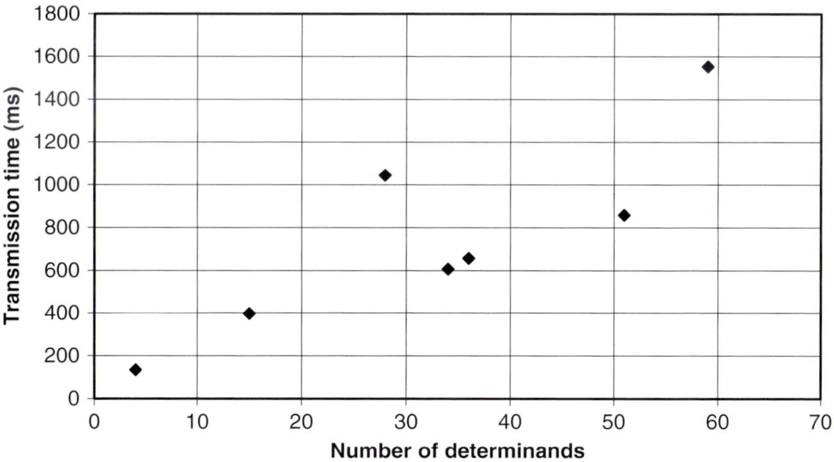

Fig. 14.2. Transmission times to user interface in determinand search case.

In Table 14.4 is given the number of determinands in each database. As can be seen, the total number is less than the sum of the individual databases, obviously because many of the determinands are present in more than one database.

When the user asks to get the available determinands, the user interface sends a simple message to the task and user agents, which then send the message to the directory agent.

Table 14.4. Number of determinands in each database

Database	IOW	NERI	UK-HMS	Waterbase	Total
No. of determinands	51	28	4	15	59

Table 14.5. Time results for search of available determinands. Times are in milliseconds

	UK	Waterbase	NERI	IOW	NERI + WB	NERI + UK + WB	NERI + UK + WB + IOW
No. of Determinands	4	15	28	51	34	36	59
Process time: TA + UA	10	8	11	8	10	8	10
Transmission time: UA-> UI	135	398	1045	859	607	657	1553
Total time (Process + transmission)	145	406	1056	867	617	665	1563
Total time per determinand	36	27	37	17	18	18	26

The reply from the directory agent is then processed by the task and user agents and the output in HTML is then finally transferred to the user interface. In Table 14.5 the processing time for the task and user agents is shown, together with the transmission time for the HTML code for different combinations of available databases. It is seen that the total time is dominated by the transmission time from the task and user agents on the main platform to the user interface on the user's PC. In Figure 14.2 is illustrated that the transmission time is nearly proportional to the number of determinants found.

14.5 THE SEARCH FOR STATIONS

The next set of queries performed in this investigation concerns the search for stations, which have data for specific determinands (the so-called Use Case 9). The number of stations found by each of the resource agents for specific determinands are listed in Table 14.6. It is seen that the number of stations differs significantly, dependent on the choice of determinand and database.

The interaction between the different agents for this kind of query is illustrated in Figure 14.3. Here is shown the output from the so-called Sniffer Agent, which is a part of the JADE (JADE, 2005) platform. Here we see that after an initialisation communication between the user agent (UA) and the DF (Yellow Pages) and the directory agent (DA),

Table 14.6. Number of stations found for different determinands

Determinand	Number of Stations				
	WB	NERI	IOW	UK	Total
Antimony			1		1
Aluminum		1	5		6
1,1,1-trichloroethane			19		19
Temperature		522		264	786
BOD		293			293
Oxygen saturation	2278	92	29		2399
Nitrate	2871	36	29	264	3200
Ammonium	3262				3262
pH	2576	506	29	265	3376

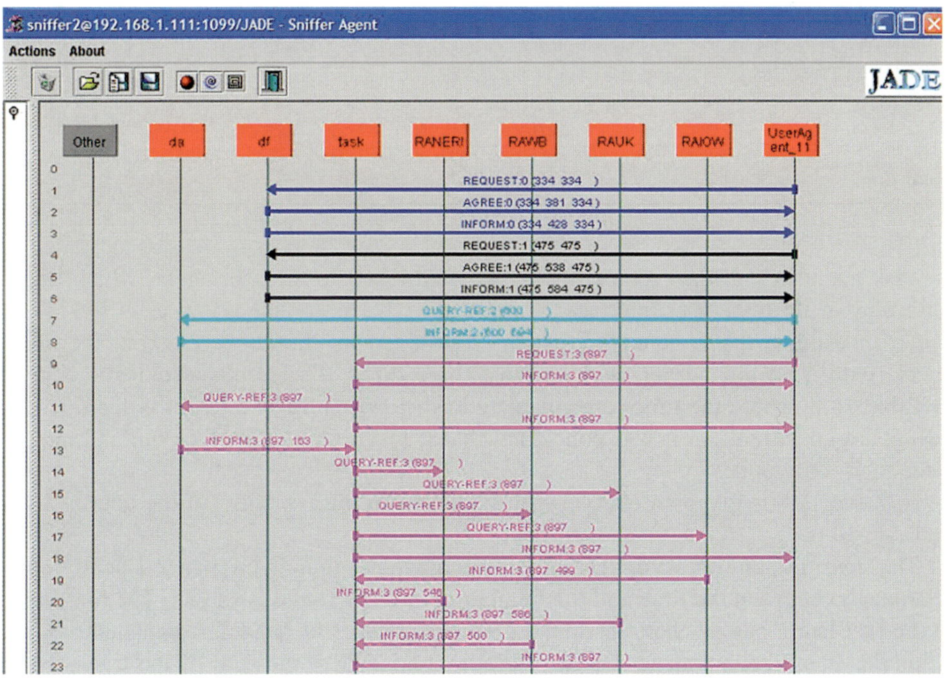

Fig. 14.3. Screendump from sniffer agent showing agent communication.

the UA sends a request to the task agent (TA), where the content is the query from the user interface (UI). The TA then sends a query to the DA to get information about which databases (i.e. resource agents) to contact. In this case, the determinand was "Nitrate", so a query to all the four RAs is sent by the TA. The replies (Inform) are then sent back to

Table 14.7. Results of a 3 times repeated query for stations. Times are in milliseconds

	Determinand		
	pH(1)	pH(2)	pH(3)
No. of stations	3 376	3 376	3 376
P:UA+TA	228	214	250
T:TA->RAWB	85	34	39
T:TA->RANERI	50	49	19
T:TA->RAIOW	16	18	46
T:TA->RAUK	23	22	29
P:RAWB	58 826	58 775	58 656
P:RANERI	27 772	27 683	27 681
P:RAIOW	1 166	1 182	1 150
P:RAUK	18 072	17 924	18 000
T:RAWB->TA	3 953	3 933	3 914
T:RANERI->TA	759	772	771
T:RAIOW->TA	48	48	50
T:RAUK->TA	531	421	541
P:TA+UA	12 122	11 534	11 606
T:UA-> UI	440	415	524
Total time	75 671	74 959	74 437

the task agent, which then returns the aggregated result to the user agent, which finally sends the result to the user interface.

To investigate the variance in the times obtained by the above method, the above query (with determinand pH) was repeated three times. The results obtained are given in Table 14.7, where the times are calculated as shown in Table 14.3. It is seen that the results are consistent, so it was concluded that it was SUFFICIENT to run and analyse each query only once.

In Table 14.8 is given an overview of the times (in milliseconds) of the queries listed in Table 14.6.

The transmission times from UA to UI are shown in Figure 14.4, where a nearly linear relationship between the time and the number of stations found is shown. The bias is due to the fact that a part of the transmitted graphics consists of headings, explanations, etc. A similar linear behaviour is shown in Figure 14.5, where the sum of the transmission times from the resource agents to the task and user agents is shown. Also the processing time required for the resource agents (and the connected MS Access database) shows a nearly linear relation between the sum of the processing times and the number of stations found (see Figure 14.6).

This relationship is not completely linear because the processing time for a given database is also dependent on the actual size (number of records). This can be seen by comparing, for example, the processing times for RANERI and RAIOW for “Nitrate”,

Table 14.8. Time results of the station queries. Times are in milliseconds

	Determinand								
	Antimony	Aluminum	1,1,1-trichloro ethane	BOD	Temperature	Oxygen saturation	Nitrate	Ammonium	pH
No. of stations	1	6	19	293	786	2 399	3 200	3 262	3 376
P:UA + TA	175	290	202	292	225	239	201	171	250
T:TA-> RAWB			13			31	99	18	39
T:TA-> RANERI		22		29	13	29	23		19
T:TA-> RAIOW	24	17				44	18		46
T:TA-> RAUK					36		26		29
P:RAWB						53 501	67 727	78 175	58 656
P:RANERI		853		16 359	28 471	5 601	2 613		27 681
P:RAIOW	401	436	818			1 242	1 146		1 150
P:RAUK					18 265		17 978		18 000
T:RAWB-> TA						3 500	4 357	5 092	3 914
T:RANERI-> TA		24		446	796	156	69		771
T:RAIOW-> TA	24	23	39			52	48		50
T:RAUK->TA					428		534		541
P:TA+UA	7 170	6 935	7 019	7 167	2 842	7 669	8 820	13 625	11 606
T:UA_> UI	148	186	118	212	266	520	443	404	524
TOTAL Time	8 324	8 321	8 242	24 517	32 625	65 471	81 711	97 598	74 437
Time pr. Station	8 324	1 387	434	84	42	27	26	30	22

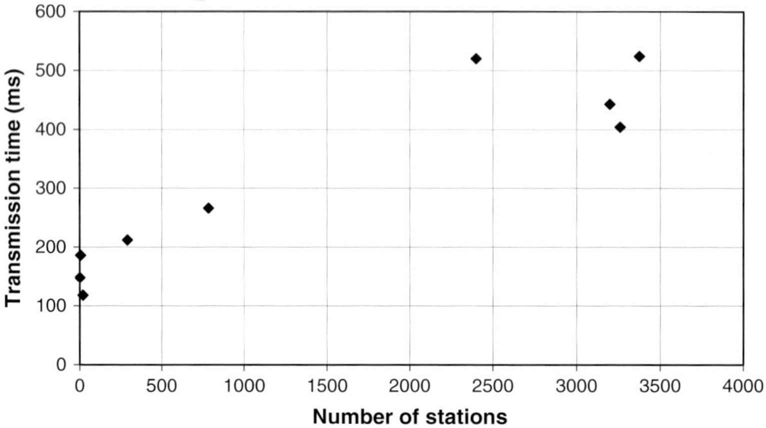

Fig. 14.4. Transmission times to the user interface in station search case.

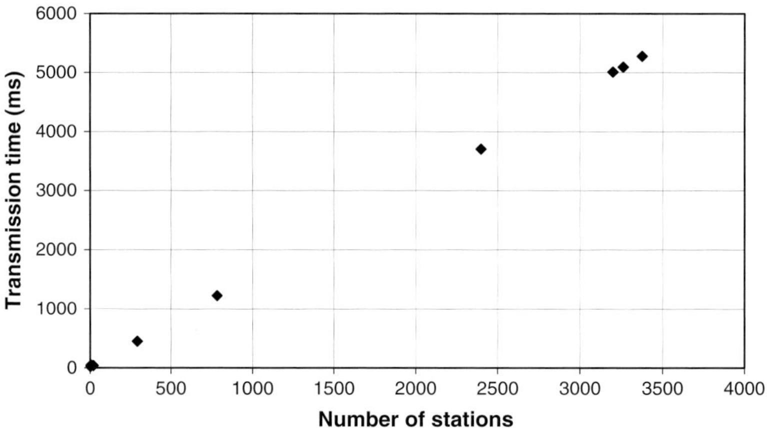

Fig. 14.5. Sum of transmission times from resource agents in station search case.

where the number of stations are more or less equal, but the processing time is only about half for RAIOW in comparison to RANERI.

By comparing the results for queries related to ammonium and nitrate, it is seen that for the same number of stations, the elapsed search time decreases when the number of databases increases. This is caused by the parallel processing by the resource agents on each platform in case of multiple databases.

In Figure 14.7 are finally shown the total elapsed search times as a function of number of stations found. Here also a linear relationship is shown, but with a certain bias, due to the (approximately) constant CPU times required to send the query and to generate the output graphics (the UA + TA and TA + UA processing times).

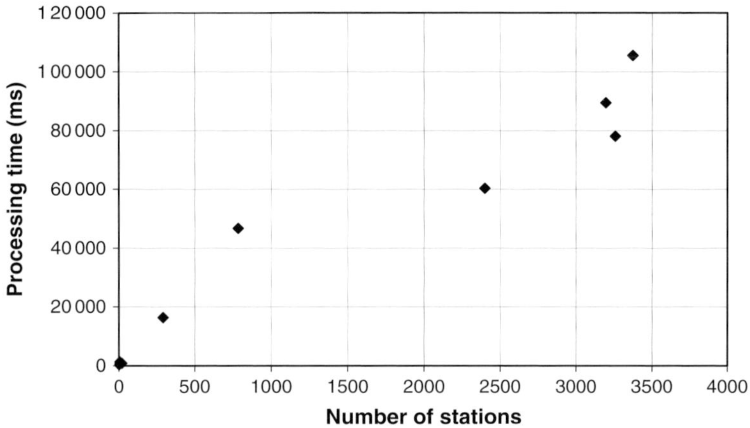

Fig. 14.6. Sum of RA and DB processing times in station search case.

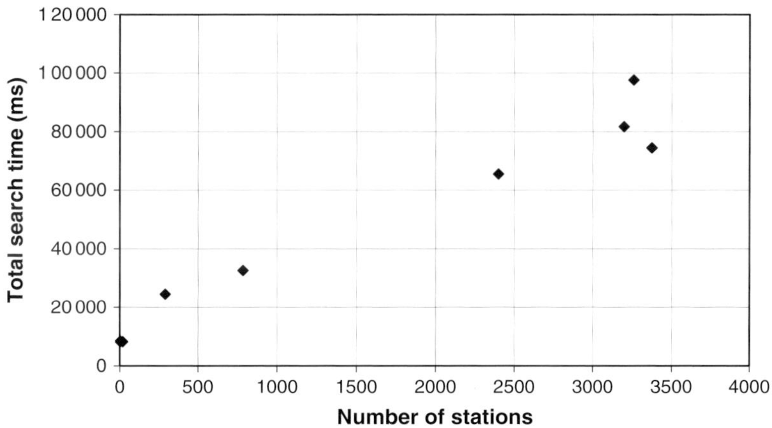

Fig. 14.7. Total search times in station search case.

14.6 THE SEARCH FOR MEASUREMENT DATA

The final set of queries available for the prototype is of the type, "What is the concentration of determinand X at station Y in the time period Z?" (the so-called Use Case 1). So in this case the databases are known and the TA queries directly the RAs, without consulting the DA. Otherwise the method applied is similar to the one outlined in section 14.3.

In Figure 14.8 is shown the output of one of the performed queries. The measurement values of the determinand pH found for four different stations in four different databases are shown here. In total, 983 measurements were made.

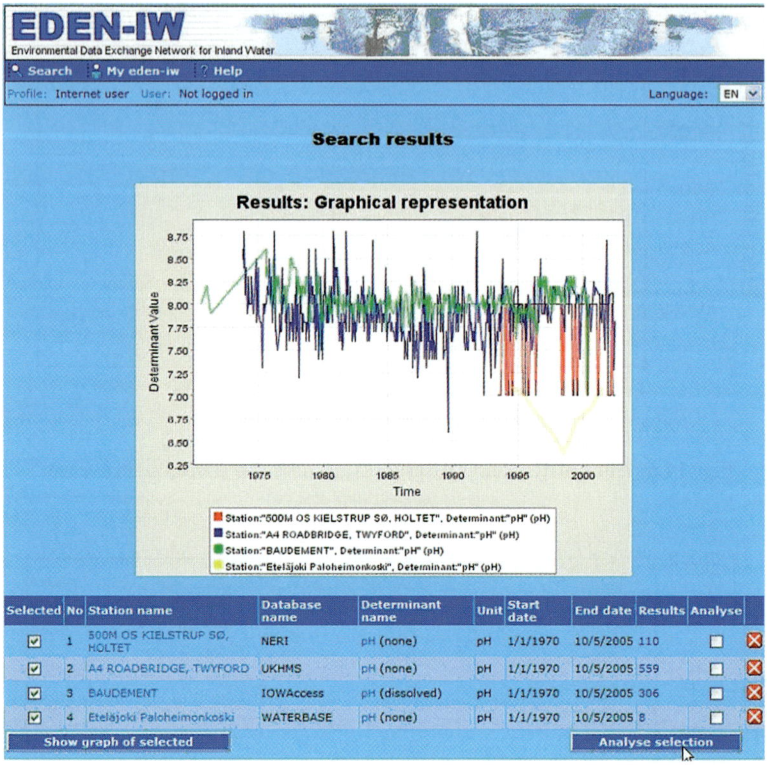

Fig. 14.8. Measurement search results from 4 databases.

In Table 14.9 is given an overview of the queries performed. It is seen that a series of measurements from 1 to 4 (or stations) have been chosen in up to four different databases.

The processing time, TA + UA, which in principle is the time for the agents to generate the graphs, is shown in Figure 14.8. The processing time is calculated as the time between the arrival time of the graph to the user interface and the arrival time of the last dataset from the RA to the TA. As seen in the table, this time varies in a non-systematic way. The reason for this is that in case of multiple graphs, the user agent sometimes generates some of the graphs in the time intervals between the arrivals of the data from the resource agents.

As in the former case, it is seen that the time per result (here measurement point) decreases when the number of results increases. Also the effect of parallel processing on the total time is illustrated by, for example, comparing the different queries for the determinand pH.

The RA + DB processing time for a given database also correlates with the number of measurements found. This can be seen in Figure 14.9. From Table 14.1, it is seen that there is only a factor of 5 between the databases in terms of number of records, so it is not

Table 14.9. Results of measurement searches

	Determinand								
	Antimony	Nitrate	Nitrate	Temperature	pH	pH	pH	pH	pH
Number of stations	1	1	4	2	1	1	2	3	4
Measurements: WB						8	8	8	8
Measurements: NERI				109			110	110	110
Measurements: IOW	1	312						306	306
Measurements: UK			1 632	395	559				559
No. of measurements	1	312	1 632	504	559	8	118	424	983
P:UA+TA	118	100	340	236	281	299	305	250	354
T:TA->RAWB						15	15	28	15
T:TA->RANERI				15					15
T:TA->RAIOW	15	15						15	16
T:TA->RAUK			69	25	16		15	16	20
P:RAWB						704	686	719	738
P:RANERI				6 796			6 892	7 029	7 019
P:RAIOW	369	17 079						16 228	16 347
P:RAUK			110 756	26 809	38 016				37 890
T:RAWB->TA						34	32	33	33
T:RANERI->TA				230			228	236	250
T:RAIOW->TA	23	653						628	627
T:RAUK->TA			3 210	850	1 110				1 111
P:TA+UA	8 856	7 440	3 506	7 407	3 374	7 860	942	7 611	1 757
T:UA->UI	14	17	28	20	27	20	35	39	337
Total	9 529	25 304	121 535	42 898	42 824	8 932	9 514	35 005	70 920
Time per measurement	9 529	81	74	85	77	1 117	81	83	72

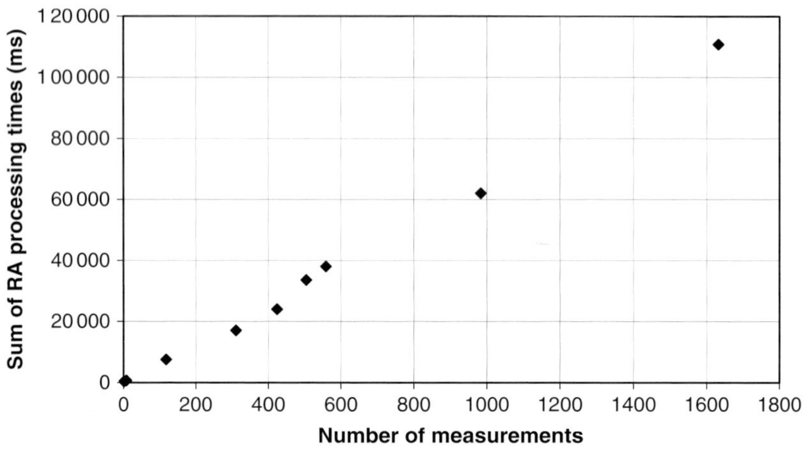

Fig. 14.9. Sum of RA and DB processing times in measurement search case.

surprising that a similar average processing time is found for the other databases. Only in the case of just a single measurement point (cf. the query for determinand "Antimony") the RA process time per measurement is significantly influenced by the number of results found.

The time per measurement is here significantly larger than that in the stations search because each result here consists of three elements: date, value and unit instead of just one–the station name.

Linear relationships, as were found in the Station Search Case above, are found in this case also. This is illustrated in Figures 14.10 and 14.11.

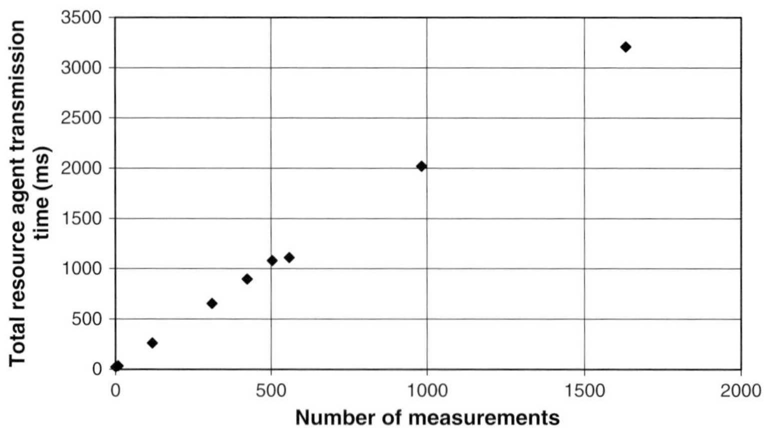

Fig. 14.10. Sum of transmission times from RA in measurement search case.

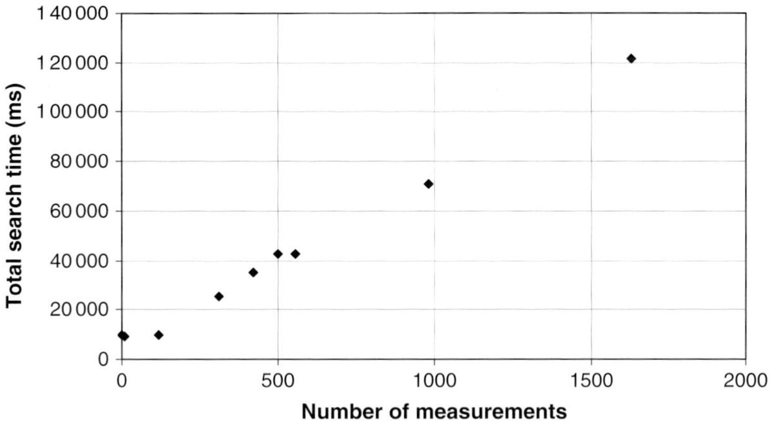

Fig. 14.11. Total elapsed search times in measurement search case.

14.7 A SIMPLE MODEL FOR THE PERFORMANCE OF A DISTRIBUTED SYSTEM

As mentioned above, the present analysis has been performed under optimal conditions with a local area wired network. In the "real world", i.e. when the databases and agents are connected via the Internet, much longer transmission times are to be expected. It should be noted that the total elapsed time for a search is dominated by the time needed for the database search. Longer transmission times are not expected to be a serious issue. As has been demonstrated, a (nearly) linear relation seems to exist between all dominant transmission and processing times and the amount of query output.

A plausible simple model of the performance of the present search capabilities in a distributed system could then be described by a linear relationship like

$$T(\text{Search}) = T(\text{Query}) + T(\text{Output}) + [S(\text{RA}) + S(\text{UA}) + S(\text{DB})]^{*} N$$

where

T(Search) is the total elapsed time for a search.
T(Query) is the transmission and process time by the TA/UA to the RA.
T(Output) is the processing time of the TA/UA to produce the graphical output.
N is the number of determinands, stations or measurements found.
S(RA) is the transmission time per N from RA to TA/UA.
S(UA) is the transmission time per N from TA/UA to UI.
S(DB) is the CPU time per N required by the RA and DB.

In the case of a determinand search, where the resource agents and database agents are not contacted, the values of S(RA) and S(DB) are evidently zero.

Table 14.10. Values of model constants for T(Search) for the wired LAN

	Determinands	Stations	Measurements
T(Query)	0	200	200
T(Output)	10	6000	6000
S(RA)	0	1.5	2
S(UA)	20	0.2	0.1
S(DB)	0	20	50

Thus, from the tables and figures above we see that in our case, with a wired LAN, the values of the constants (in milliseconds) are approximately as in Table 14.10.

As mentioned above, these values are expected to differ in an Internet-distributed system. Here the values of T(Query), S(RA) and S(UA) are expected to increase due to the higher transmission times on the Internet.

The CPU time for the database search (S(DB)) consists of the CPU time for the RA and for the database itself (in our case, MS-Access). In order to estimate the distribution between these, a series of direct searches in MS-Access were made with the same criteria as given in Tables 14.8 and 14.9. In all cases, these search times were less than 1 CPU second, so it can be concluded that S(DB) is dominated by the CPU time for the RA. Therefore, an optimisation of the database itself would only give a marginal reduction in the S(DB) value.

An attempt to extrapolate the above results to an Internet distributed system was made by estimating the download transmission times to the JRC in Italy from the three home countries of the databases. Here, the Web-speed measuring facilities at the websites listed in Table 14.11 were applied.

In the above table, also the speed from the US was estimated to model the case with a database located, for example, at US-EPA. The main platform could be located in any of the listed EU countries.

In the actual setup, the transmission speed was 10 Mbps. From the above table it is seen that an average download speed in the order of 500 kbps could be expected in a real configuration. Thus, a reduction in the transmission speed with a factor of 20 seems

Table 14.11. Download transmission speeds from different countries to the JRC

Country	Website	No. of tests	Average transmission time (s)	Average transmission speed (kbps)
France	www.numion.com	25	60	600
UK	www.numion.com	25	60	400
Denmark	www.gemal.dk	25	30	1000
US	www.numion.com	25	60	300

Table 14.12. Estimated model constants for T(search) for connections via the Internet

	Determinands	Stations	Measurements
T(Query)	0	4000	4000
T(Output)	10	6000	6000
S(RA)	0	30	40
S(UA)	400	4	2
S(DB)	0	20	50

Table 14.13. Estimated search times for connections via the Internet

	Determinands		Stations		Measurements	
N	5	50	20	2000	10	1000
T(s)	2	20	11	118	11	102

realistic. If we maintain the CPU times for the database search (S(DB)), we get the constants listed in Table 14.12.

Here, we have made the conservative assumption for T(Query) that the process time can be neglected in comparison to the transmission time.

From the above constants we would expect the search times listed in Table 14.13 for the three Use Cases (for two representative number of results).

From the table, it is seen that, despite the large decrease in transmission speed (a factor of 20), reasonable search times (2 minutes or less) can be expected for realistic searches on an Internet distributed system.

14.8 SUMMARY AND CONCLUSIONS

An investigation of the performance characteristics of the EDEN-IW prototype was made on a wired LAN.

Initially, a wireless 11 Mbps hub was used, but due to technical problems this approach was abandoned. It was found that sometimes the wireless connection caused the agents to hang or even exit during a query. In many cases, it was necessary to restart the agent platform(s) to remediate the problems. Sometimes it was also necessary to restart the main agent platform to reset the prototype.

Similar problems were experienced in previous tests when the agent platforms were connected via the Internet. However, it should be stressed that the EDEN-IW Java agent platforms are based on a version of JADE (version 2.5), which other developers also found to have the same kind of difficulties. An improved version (version 3.3) of the JADE framework is now available, where these problems should be solved.

The investigation revealed that the major part of the time required from the time when a user submits a request until the result is displayed at the user interface is taken by two different processes:

1. The database search done by the database (here MS Access) and mainly by the resource agent
2. The result-presentation process done by the task agent and mainly by the user agent

For larger databases, the database search process seems to be the most time-consuming part, as it increases more or less linearly with the database and result size, whereas the presentation process time is more or less constant.

Only in the search for available determinands, where the process is a simple table lookup at the directory agent, the time to transfer the messages between the agent platforms forms a major part of the total search time.

A simple model for estimating the total elapsed query time was proposed and quantitative values for the constants in the model were given for the present investigation, which was performed on a wired LAN.

An extrapolation to a "real world" situation, where the databases are distributed in their home countries and connected via the Internet, was attempted. Here the transmission times play an increasing role as the transmission speed is reduced by a factor of 20. This exercise revealed that, despite the large observed decrease in transmission speed, reasonable search times (2 minutes or less) could be expected for realistic searches.

Thus, it can be concluded that the main focus for further investigations to enhance the performance of the EDEN-IW prototype should be on improving the reliability and stability of the agent system in realistic networks connected via the Internet.

PART III

PERSPECTIVES AND DATA HARMONISATION

Chapter 15

ENVIRONMENTAL DATA: EDGE ISSUES AND THE PATH FORWARD

B. Bargmeyer and S. Chance

15.1 INTRODUCTION

In Part II of the book, it was illustrated that EDEN-IW succeeded in demonstrating that it is possible to use systems of agents to access environmental data. In this case, it concerned inland water data via ontology-based queries sent to independent databases housed on heterogeneous platforms. The queries could be sent to disperse databases and the results could be brought together, merged, displayed, and used for a wide variety of environmental action and decision-making. The core of the success is the interaction between agents, brokers, and mediators – to formulate the query, send the query to potential sources, retrieve the results, transform the results from different sources into a common form, merge the transformed results, and present the result to the user. At the edge there are challenges, difficulties, and a major area, semantic computing, where continuing research development and demonstration is needed.

Some longstanding challenges were naturally excluded from the focus of the EDEN-IW research. These remain as edge issues that constrain the application of agent technologies. These include query optimization, user interfaces, and the effect of firewalls. However, at the leading edge, emerging technologies offer promise. These include: Semantics management in an eXtended Metadata Registry (XMDR), Ontology Lifecycle Management (OLM), and Semantics Service Oriented Architecture (SSOA) and semantic grid services.

In the following sections, a number of these issues will be discussed starting from some of the immediate operational issues of query optimization, through to firewalls, then into a major R&D challenge: Semantic Computing.

15.2 OPERATIONAL EDGE ISSUES

15.2.1 Query Optimization

EDEN-IW avoided situations where query optimization would have a critical impact on system operation. Query optimization for ad hoc queries is a challenge for large databases, even on centralized systems. It gets harder as data is distributed in a homogenous platform environment and gets progressively harder as data is accessed across dispersed, heterogeneous platforms. Even implementation of the query optimization techniques found in any single vendor's DBMS product would significantly increase the complexity of agents and their interactions. In a related problem, many data sources are loath to provide machine space and computing cycles for storing intermediary results, such as table joins for further processing. While computer memory usage and CPU time may be acceptable costs to support users within an organization, it may not be acceptable to cover such costs for external users, which may draw on substantial system resources to process the intermediate results of complex queries. Finally, the notion of browsing large ontologies, selecting random data concepts to query, and sending off queries to unknown resources is fraught with the likelihood of unacceptably long response times, especially given the impatience of Internet users. The EDEN-IW project exposed this issue. However, for EDEN-IW, the use of pre-designed menus of parameterized queries (with acceptable response times) sidestepped this challenge and enabled the limited resources to be used for the core challenges of an agent-based system.

15.2.2 User Interfaces

Developing user interfaces, especially culturally adaptable interfaces, can absorb a huge amount of time and funds. Design criteria for workable, menu-driven interfaces are well known and adequate interfaces of moderate complexity can be developed as straightforward projects that may nonetheless be expensive. Sophisticated user interfaces are the subject of R&D, but were not part of the EDEN-IW project. Interfaces for browsing ontologies and formulating ontology-based queries are a large challenge. Other projects have developed some fascinating interfaces to display and browse the nodes and relationships that can be found in ontologies. But the fascination fades as we engage in real tasks, where we must find desired data in a large, unknown concept space. As the number of concepts in ontologies increases from 10's to hundreds of thousands and the number of terms and relationships passes millions, simple interfaces fail. Again, for EDEN-IW, the use of pre-designed menus of parameterized queries was the most achievable path forward, given limited resources.

15.2.3 Firewalls

An issue that arose strongly as the project proceeded was the crushing effect of firewalls and system security. Ports such as the one for accessing SQL are being slammed shut as networks and operating systems go from "open" by default (unless there is a need to restrict access) to "closed" by default, with access granted on a sparing basis. This extends to running only "approved" software in production environments, which makes the introduction of new software difficult. It leads to issues about interacting with external agents and brokers. Thus, the security model for agent-based computing becomes much more complex. As a consequence some of the easier achievable results are denied by security and firewall mechanisms.

15.3 THE SEMANTICS EDGE: A MAJOR CHALLENGE AND PROMISING OPPORTUNITY

Semantics is at the core of EDEN-IW. The agents, brokers, and mediators are mechanisms for accessing, retrieving, and manipulating data. However, the essence of EDEN-IW is accessing data that is generated and managed by independent efforts. When the agents, brokers, and mediators all work, we are faced with the data content, particularly the semantics of the data, and dealing with that is the next major challenge.

While dispersed databases may have comparable data, the same facts (concepts) may be represented with different terms, different codes, different measurement units, etc. For example, the country Italy can be represented according to ISO 3166 – *Codes for the representation of names of countries and their subdivisions* in several ways. The short English name is "Italy", the three-character alpha code is "ITA", the two-character alpha code is "IT", and the three-character numeric code is "380". Similarly, quantities may be measured in units that scale by a factor of 10 (metric, e.g., liters, centiliters, deciliters) or in English units, etc. There are thus several ways to represent the same observation or "fact" and there is no hope of reducing this to a single representation across all data sources. As query results are returned from the dispersed databases, most users will want the data to be transformed to a common representation with a coherent set of codes, units of measures, consistent terms, etc. Part of the process of results integration must be to transform the results into a representation favoured by the user.

EDEN-IW makes clear the need for a semantic mapping service which would take in one representation of a concept and output the same fact in a different representation. The input might be "ITL" and the output could be "Italy". A similar service is needed to convert data to a common unit of measure. The EDEN-IW project developed modules to do this for certain known transformations; however, it exposed this area as one needing a more fundamental and general approach. Metadata registries are a promising technology for providing semantic mapping.

Another semantic challenge is to establish, for some community of interest, an agreed-upon set of concepts, concept definitions, and associated terms. Classic terminology would define a concept by establishing the essence and differentia for the concept. However, concepts can also be identified through their relationships to other concepts. For example, a concept (say, concept A) may evidence specific relationships to other concepts (say, concepts C, D, E, F, G). If another concept (say, concept B) is related to C, D, E, F, and G in exactly the same ways as is concept A, then concepts A and B may be the same. At least this might assist a human to find correspondences between large concept systems.

Agents are a foundational technology for the semantic web. Agents can act on information stored as RDF triples. RDF triples (subject, predicate, and object) require a "vocabulary" that establishes the meaning of the terms used for the subject, predicate, and object. Here again, the agent can only proceed reliably if there is an agreement between some community of interest as to the meaning of the concepts. While agent systems can work without Web technologies such as XML and RDF, and without semantics management, there is much to be gained by convergence of agent, semantic web, and metadata registry technologies.

The EDEN-IW results bring use cases and requirements to an emerging field of semantics management and semantic computing. Other use cases and requirements are flowing in from many directions to focus new R&D efforts in this area. The following sections will discuss major research needed in the areas of metadata registries, ontology lifecycle management, and semantic service oriented architecture.

The basic thrust is to R&D the use of metadata registries to:

- Register and manage any semantic artefacts useful for data management/data administration
- Link concept systems to data
- Ground RDF statements and ontologies
- Provide ontology lifecycle management associated with data lifecycle management
- Provide semantics services in a semantics service oriented architecture.

15.3.1 Semantics Management and Data Administration

Many organizations have spent substantial resources to standardize data and to integrate database schemas to achieve system interoperability and other goals. These efforts did produce substantial accomplishments. However, the approaches often could not be sustained, due to the inability to scale to cover the broad inter-organizational datascape and due to advancements in other areas of computing that outran the methods being used. Metadata registries, such as specified in the ISO/IEC 11179 family of standards, provide substantial support for this endeavour. The underlying challenge continues to grow as the range of data trading partners increases and as more recent technologies such as XML data transfers and XML schemas gain usage.

Advances toward semantic computing make it useful to link the data element metadata (definitions, valid values, etc.) to the more fundamental semantics found in concept systems and Knowledge Organization Systems (KOS). The concept systems and KOSs may vary in size from hundreds of concepts to hundreds of thousands or millions of concepts, along with millions of terms and millions of relationships between concepts. The concept systems may have no relationships between concepts or may have a variety of relationship types (broader-than, narrower than, is-a, has-a, . . .). The concept systems may have simple or complex structures. This could be directed acyclic graphs, undirected cyclic graphs, faceted classifications, lattices, bipartite graphs, directed graphs, cliques, etc.

The challenge here is to find the means to register and manage the semantics of:

- Data elements, data element concepts, value domains, and other metadata associated with data systems and XML schemas
- Concept systems and knowledge organization systems, including keywords, thesauri, taxonomies, ontologies, and axiomatized ontologies
- Relationships between the data elements and concepts
- Relationships between the concepts themselves.

An additional challenge is to provide services that make the above useful for designing and maintaining databases, for XML data transfers, for agent-based computing, and for semantic computing such as the semantic web.

A focus of the research suggested here could be to extend semantics management and data administration capabilities of metadata registries beyond that supported by the current metadata registries to create an eXtended Metadata Registry (XMDR) to address more fundamental issues that arise in semantics-based computing.

Linking semantic artefacts to data

There is a need to link items found in text, voice, and other sources to data that might be relevant. Individual x, disease agent y, and place z might be found and extracted from some medical records. There is a need to link the extracted concepts to terms in concept systems and to data held in databases about individuals, genetics, disease agents, and places. This could, for example, relate to cancer research. Other applications include bio-terrorism or eco-terrorism.

In the US, the National Institutes of Health including the National Cancer Institute have developed several thesauri in this area. They are now in the process of transforming them into ontologies, some using formal description logic. While NIH and NCI may use a wide variety of software to create and maintain the ontologies, it may not be practical for organizations to access these (and many other) sources directly. For many practical reasons it would be useful to gather the concept systems (e.g. the ontologies) together in one place, possibly with versions of each.

A R&D project could explore the challenges involved in extending metadata registries for the purpose of registering ontologies, registering relationships between them, and registering relationships between the concepts and data in databases. This could be extended to Ontology Lifecycle Management and to the provision of semantic services.

This work would draw together three meshes of thought and action:

- Semantic mark-up and inference systems (DAML, RDF, OWL, Semantic Web, . . .)
- Metadata registries, and
- Semantic content (e.g., thesauri, taxonomies, and ontologies).

Semantic markup and computing

R&D projects have created technologies that enable software agents to dynamically identify and understand information sources, as well as technologies that provide interoperability between agents in a semantic manner. These technologies dramatically improve the utility of Web-based queries and services. A vision for the semantic web was set forth in an article by Berners-Lee, Hendler, and Lassila in the *Scientific American* in May 2001. This article raised the possibility that a browser query and software agents could produce a specific answer to a specific question, ridding the user from having to plough through hundreds or thousands of responses of which only a few may pertain to the area of interest. In order to do this, the relationships between objects and processes have to be readily identified and linked, thus, greatly increasing the computer's capability to evaluate data.

Mark-up languages enable capture of the semantics of data and automation that can leverage semantics to facilitate alignment of information.

Essential to this endeavour, but beyond the scope of the initial R&D projects, is the creation of the large-scale semantic content upon which the semantic-based tools and agents could operate. However, an essential requirement of semantics computing is a "vocabulary". For example, the nodes and edges found in RDF triples have URIs that point to references for the terms used as the subject, predicate, and object. While RDF allows the URIs to point to nothing, this seriously detracts from agents' abilities to deal with the semantics and related issues such as trust. Vocabularies are essential, preferably ones vetted by some community of interest, such that the semantics can be trusted for a particular purpose. Connecting the emerging semantic techniques and technologies to semantic content of known provenance should be a major focus of R&D for extending metadata registries.

Semantic content

There is a long tradition of developing semantic content for specific organizations or communities of interest. This includes controlled vocabularies, thesauri, and taxonomies in areas such as the environment, healthcare, biology, transportation, chemistry, etc. Even ontologies are beginning to appear in some of these areas. Initial uses of this semantic content have been for functions such as glossary/dictionary (what does a term mean?), library cataloguing (what subjects does this document address?), thesauri (what terms are broader, narrower, and related?), taxonomic (where does something fall in a hierarchy?), etc.

While many of these efforts began as ways to define and manage "terms", some of this work has shifted to the management of "concepts", which may be given linguistic expression in one or more languages. To some extent this work has been coupled to data administration, wherein concepts (terms) may be organized into data elements, which may be expressed as database schemas or XML schemas, along with valid values in a "value domain", wherein each value is a concept expressed as terms in some language or encoding. Metadata registries are a good environment in which to pursue such coupling of the fundamental semantics of concepts to the semantics of data.

There is an increasing need to be able to draw together content expressed in various structures, to interrelate the concepts found in the various structures, and to link the concepts to data held in databases. Metadata registries could provide a useful instrument for these capabilities. While the concept systems may be created and maintained in a wide variety of systems, it should be possible to register and interrelate the contents in a metadata registry.

15.3.2 Metadata Registries

Governments have supported R&D and standards development in the area of metadata registries to manage the semantics of data. Initially driven by the need to derive comprehensive information from queries against disparate, stovepipe computer systems, the emphasis has shifted to achieving semantic consistency in XML data interchange, and now is evolving to the management of the semantics underlying data structures. The results of this work progresses through standards committees such as ISO and IEC to become, e.g., the ISO/IEC 11179 – Metadata Registry family of standards.[1]

Government agencies such as the European Environment Agency, the US Environmental Protection Agency, the National Cancer Institute, and others have invested millions of dollars building ISO/IEC 11179 metadata registries as facilities in which to register data semantics, such as the definitions for data elements and the enumerated values (with definitions) for the permissible values of data elements.[2] Large sums have been spent on developing and registering concepts, definitions, and other descriptive information in various semantic structures. In addition, the agencies continue to extend and maintain the semantic content specifications through interaction with communities of interest in the environment, health, socio-demographics, transportation, space science, and other areas. The semantics in these registries have often been vetted with broad agency constituencies. The content of these metadata registries, if made available through appropriate services, would do a great deal to bootstrap the semantic web and to provide the semantics needed for semantics-based computing.

Advancements in metadata registry technologies are required to enable the registration of semantic diversity, the linking of semantic correspondences (e.g. across languages, dialects, and disciplines) and the tracking of semantic drift as languages, terms, and concepts evolve. This would enable communities of interest to converge on consistent use of concepts and terms, by supporting harmonization efforts. In addition to other applications in the area of intelligence, harmonized concepts and definitions will help in the formulation of legislation, regulations, programs, and information systems.

Note that harmonization does not imply the selection of a single term or code to represent a concept. For example, the terms "typhoon", "cyclone", and "hurricane" are likely to continue to refer to a similar weather phenomenon in different geographic contexts. Similarly, a single concept will find linguistic expression in multiple languages and multiple dialects (or regionalizations) of the same language.

A primary distinction of the evolving technology should be its capability to deal with diversity and concept drift, as opposed to notions of "fixing" concepts and terms in a static condition (and using technology limited to such capabilities). This is a level of complexity that current technology does not well address.

An exemplar and vital service for semantics-based computing could be a metadata registry exposed as a semantic service. It would facilitate automatic discovery of relevant ontologies or other artefacts required for broad applications. The availability of metadata and/or ontologies in a centralized mechanism promotes ontology sharing, reuse, merging, and alignment. Each of these facilitates wide-scale implementation of semantics-based computing applications. Software agents will be better able to interact with numerous ontologies to retrieve applicable information. Secure computing environments raise the requirement for gathering concept systems into a common store, since the concept systems found in the wild are not directly accessible within a secure environment.

When emphasis is placed on SBD and business logic is separated from implementation technology, the IT return on investment (ROI) is substantially increased. Indeed, the application's logic provides value to an organization; technology is *how* business logic is actualized. Specifying business logic in a technology-agnostic mechanism promotes stability and agility. An organization less susceptible to disruptive technology is more adaptive to change, and thus more likely to succeed.

Standard Based Design allows rapid adoption of emerging technologies, technology-independent representation of application logic, and application life cycle support. Semantic Web constructs and technologies bring logic, context, and automation. It is the combined effect of SOA, Semantic Web, and SBD that forms the foundation of SSOA – and other emerging standards-based, semantic, service-oriented software implementations.

Semantic Web

As mentioned earlier, the Semantic Web is composed of three basic components: Knowledge Representation (KR) languages, ontologies, and logic. Although not a primary Semantic Web component, agent technology is an important beneficiary of the Semantic Web environment. The real power of the Semantic Web will be realized when these are operating seamlessly and synergistically in concert.

Knowledge representation captures semantics of data, enables automation that can leverage semantics to facilitate alignment of information models, and fulfils a prerequisite to information interchange among multiple actors. Standardization of KR languages is a significant step in ensuring the Semantic Web will become a reality. The Extensible Markup Language (XML) and Resource Definition Framework (RDF), now well established with significant implementation experience, provide the baseline for the Web Ontology Language (OWL), the W3C Recommendation for a standard KR language. OWL provides the abstract syntax enabling content to be tagged with semantic meaning by describing it relative to other described concepts and establishing relationships between them. OWL is only one standard for KR; others, such as Simple Common Logic (SCL), exist. In fact, SCL is considered more expressive than OWL; however, OWL is the standard expected to achieve widespread implementation across the Internet. Formal ontologies ensure content is described unambiguously, enabling interpretation by capable machines. It is important to note that through the SSOA services, employing OWL will interoperate with SCL or other languages as the agreement will be in the interface.

A portion of the logic component of the Semantic Web is partially embedded within the ontology; additional logic can be applied by incorporating external, situational, or event-driven rules. The logic embedded within the standardized KR language and the ontology, combined with optional external rules, forms the domain theory. The domain theory governs the actions of all activities within the domain. Logic affords machines and software agents the ability to reason against the described content to answer questions, classify concepts, and fire rules triggering action. Once this component is established, machines can negotiate, render conclusions based on an interpreted set of facts, or optimize functions based on constraints. These actions will be accomplished with little or no human interaction, freeing the human to focus on the tasks requiring higher order reasoning.

Agents, empowered by the semantically rich environment, can be used to accomplish the mundane, repetitive, and time-consuming tasks currently occupying the majority of human analytic time. For example, E-commerce transactions will rise to a new level of efficiency as mobile, autonomous agents transit between vendors to accomplish Business-to-Business (B2B) or Business-to-Consumer (B2C) tasks on behalf of human actors. Agents can continue their tasks indefinitely as they have unlimited endurance, yet still require a human user to specify parameters regulating their actions and behaviours. Agents will be capable of communicating with other agents to pass value added information forming societies of agents.

Semantic Service Oriented Architecture may be derived from SOA, SBD, and Semantic Web concepts. SSOA will ultimately allow goal-based business operations to occur. That is, business analysts will submit a query or require some task completion and the workflow will be automatically created and dynamically executed based on available semantic services. Service providers will define standard interfaces that will be exposed for eligible service requestors to invoke. Services will be automatically discovered and consumed based on analysts' business requirements.

The Meta- Factor*

Implementing SSOA concepts yields an empowering capability. Semantics-based computing *begins* with highly structured data that represents data elements in a machine-interpretable manner. Ontology artefacts may be created and registered in a Semantic Metadata Registry. Metadata may be reposed in a shareable, persistent state for other users (human or agent) to reuse, extend, or align with. Sharing not only enables reuse, it also promotes enterprise data element standardization. This results in a more common awareness and understanding and facilitates policy definition and compliance.

Semantic metadata concepts may be extended to services. For example, a Semantic Meta-Service Registry could provide a similar capability as a Semantic Metadata Registry. For example, semantic service metadata could be stored in a repository to enabling more sophisticated service management. Relationships among registered items could be discovered, facilitating specification of an optimized Semantic Service Taxonomy. Enterprise services could be refined to reduce or eliminate unnecessary redundancies. Service design patterns could be enhanced due to information gleaned from a Semantic Meta-Service Registry.

Continuing, a Semantic Meta-Process Registry may provide a mechanism for business process reuse. Business analysts who create end-to-end processes for agents to execute can register created processes along with process descriptions in a registry for sharing and reuse. Obviously, this situation enables standardization of business processes. Process extension and generalization are facilitated. Process extension allows other users to retrieve registered processes and add functionality, or "connect" to other processes. Process generalization enables "process patterns" to be reused and newly created processes to be constructed based on existing processes. This not only contributes to business process alignment, it speeds execution of business goals. Best practices are readily discernible and describable and may feedback for training of new analysts.

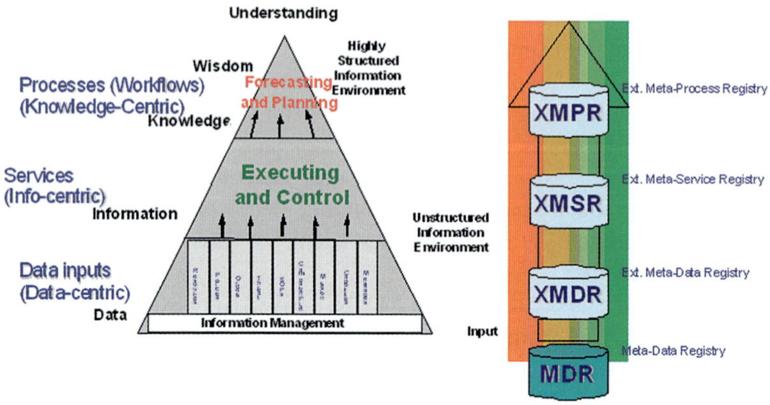

Fig. 15.1. The Meta-* factor: Data to services to processes.

These three Meta-* levels combine to help reverse the current state of IT-driven organizations to goal-oriented enterprises (Figure 15.1). Instead of concentrating on solving "impedance mismatches" between static applications, business analysts focus on higher order reasoning tasks. IT (finally) serves the human instead of the human trying to satisfy IT.

15.4 CONCLUDING REMARKS

The maturation and adoption of semantics on agent technology will change the roles of computers as we know them. Agents, empowered by a semantically rich environment, will be used to accomplish mundane, repetitive, and time-consuming tasks currently occupying the majority of human computing time. For example, E-commerce transactions will rise to a new level of efficiency as mobile, autonomous agents transit between vendors to accomplish Business-to-Business (B2B) or Business-to-Consumer (B2C) tasks on behalf of human actors. Agents can continue their tasks indefinitely as they have unlimited endurance, yet still require a human user to input parameters regulating their actions and behaviours. Agents will be capable of communicating with other agents to pass value added information forming systems (i.e. societies) of agents.

The disciplines concerned with eliciting and codifying the knowledge contained in the ontology are Knowledge Acquisition and Knowledge Representation. Knowledge Acquisition and Knowledge Representation are usually done by a Knowledge Engineer in cooperation with one or more domain experts. Semantics-based computing will strive to make these critical disciplines accessible to the novice; however, the effects of complexity must be overcome in these disciplines first. Knowledge Representation specifically is the application of logic and ontology to the task of constructing computable models for some domain (Sowa, 1999). Knowledge Acquisition is the description and exposure of domain expertise. As mentioned above, when an object is described, it becomes more complex. Domain knowledge as it is represented for inclusion into an ontology for use in an application, also becomes more complex for the same reasons all traced to the property of complexity.

NOTES

[1] The ISO/IEC 11179 family of standards includes ISO/IEC 11179 Parts 1–6 as well as multipart standards ISO/IEC 20943 and ISO/IEC 20944. These can be obtained for free at www.metadata-standards.org. Revisions are underway to result in a new edition of 11179. Drafts of proposed revisions are also available at that website.

[2] See the US Environmental Protection Agency's Environmental Data Registry at www.epa.gov/edr and the National Cancer Institute's Cancer Data Standards Repository at ncicb.nci.nih.gov/core/caDSR.

[3] "Seamlessly" indicates that participating services are made available using well-known, standard interfaces; thus precluding implementation of multiple, ad hoc standards that result in brittle, inefficient applications.

Chapter 16

A WATER REFERENCE FOR EUROPE

A. de Jager, E. Rimaviciute and P. Haastrup

16.1 INTRODUCTION

The harmonization of environmental policies in the European Union regarding the protection of water supplies is, since 2002, focusing on the so-called Water Framework Directive (WFD) (EU Parliament & Council, 2000). This directive states as one of its goals the ambition to enhance the exchange of geographic data describing the location of objects relevant for the monitoring of water quantity and quality. This chapter will focus on the creation of this common European Water Data Reference System in which geographic data ranging from fresh water bodies like small ponds to tidal areas along the coast are to be shared between the member states committed to this directive[1]. All the objects in the reference are to be monitored regularly and consequently classified.

Water bodies classified as performing poorly with respect to their ecological status will be subject to specific measures of the Union, thus allowing authorities to assure clean and sufficient water resources by 2015.

Europe has heterogeneous and rich geographical data sources and therefore risks to obtain a geographical reference of varied quality, containing extensive information in one area but too poor information in other, perhaps bordering, areas.

The Figure 16.1 below gives an example on the indexed heterogeneity of data delivered in the initial build-up of a catalogue of main rivers of the WFD. As a start to build a geometric reference, 19 member states already submitted datasets, varying both in geometric quality and in coverage of their respective territories. The index displayed in this map shows the difference in density of the rivers provided.

For decision-makers at continental level, information of comparable quality is highly desired in order to assess the progress of the implementation of the directive. Furthermore, the water resources of the European Territory are influenced by states not participating in the directive, like the Western Balkan countries, Moldova, Belarus, Russia and the Ukraine.

In order to overcome these hindrances, methods were developed to create a geometric reference using data originating from space (e.g. satellite images), called space born data in the following text, and various data enriching modelling methods.

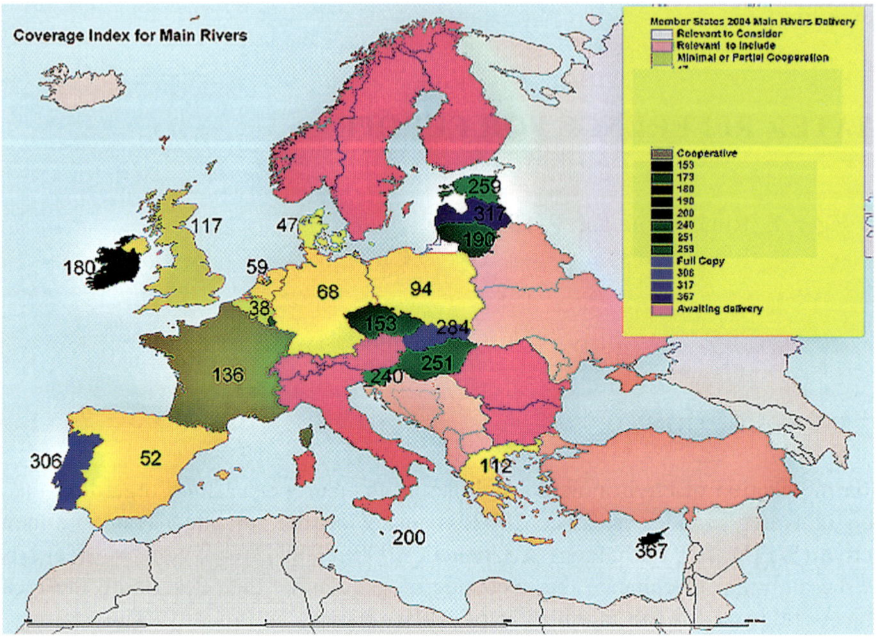

Fig. 16.1. 19 member states already submitted datasets.

Space data allows both for detection of surface water bodies and for the detection of height of the surface. Space born data can also be used, e.g., to detect oil slicks or sudden growth of algae in the summer season. Although these methods are also very promising as verification methods within the scope of the WFD, this elaboration focuses on the application of space born data in the context of the creation of a geometric reference data only.

The main geographic reference data to be created are water bodies like streams, channels, networks of channels, lakes, canals, tidal areas and swamps. Furthermore, one has to be able to represent these objects at various scales, ranging from 1:25.000 to a Continental map displaying just the major watersheds in a specific colour. In total, it is expected that the reference will cover over half a million of data elements (records), covering an area from the Azores Islands to the Ural and from the North Cape to Cyprus. This is shown in Figure 16.2. The colour in the figure indicates the number of countries the basin is situated in. The green basins are basins that are situated in only one European State. The Danube (in orange) covering 13 countries is the most complex and extended system. The grey boxes are 1 decimal degree tiles that correspond to the SRTM tiling system. In total, the European Area of Interest covers 2180 Tiles, covering the Asian part of Turkey and Russia up to the Ural Mountains and Caspian Sea.

In order to facilitate data exchange regarding the various water bodies and their respective quality status, a methodology for coding the various objects is proposed.

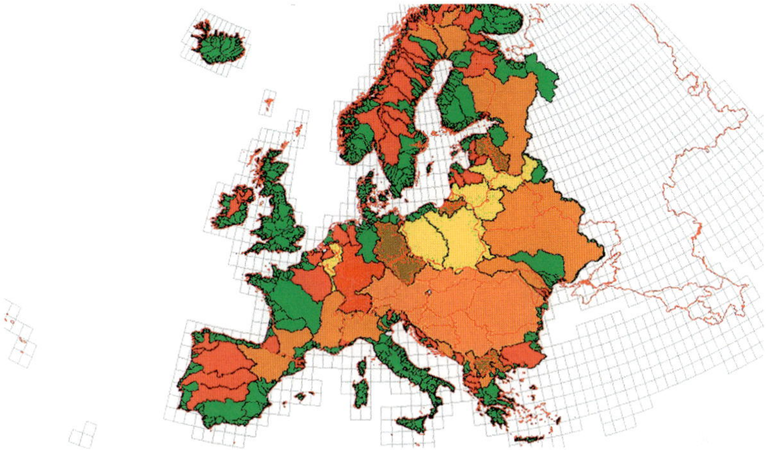

Fig. 16.2. Overview of the watersheds of Europe as generated with the CCM version 1.

With regard to coding many requirements exist and have been described already in various requirement documents. It is however important to develop a human-readable code that can help with data recognition and validation, but still can be largely derived using machine-made algorithms.

16.2 COMPONENTS OF THE REFERENCE

The geometric reference will consist of various components, some of which can be detected automatically using space born data, others by using existing or upgraded catalogues in use by the various authorities in charge of preserving water quality or quantity.

The basic building block of the reference will be the watershed having a specific outlet in the sea.

Europe like any continent can be subdivided into a relatively small number of major watersheds that cover most of its area. Commonly known are the Danube, the Rhone and the Rhine basins. In the WFD terminology, these areas are referred to as River Basin Districts and they may already be managed by umbrellas of supra national water authorities with the mandate to handle the common interests of the extended areas that share the same freshwater flow.

These major watersheds are built up of catchments, each containing segments of the river that make up a leaf of interconnected water discharge channels of the land.

Of course, lakes and groundwater bodies also make up a substantial part of the reference. These can be connected to the rivers in the watersheds – and in exceptional cases even cross them.

Table 16.1 gives an overview of the number of elements that can be expected for the catalogue to be developed. The amount of data anticipated in this overview relates

Table 16.1. Overview of the main elements of a water data reference

Element	Expected number	Subdivision of
Continental Seas	25	Ocean (2)
Continent and Islands	2 000	
Watersheds	8 000	Continent and Islands
Catchments	1 00 000	Watersheds
Rivers	30 000	Watersheds
River segments	4 00 000	Rivers
Lakes and Marshes	1 00 000	Watersheds
Canals	80 000	Land and Islands
Dams	4 000	Lakes
Groundwater extraction	1 00 000	Land and Islands

to a nominal map scale of 1:100.000. Most elements can be represented as polygons. However, for operational purpose it is often desirable to register rivers, river segments and canals as line elements in which the draw direction of the line corresponds with the natural flow of water. Europe is surrounded by two oceans: the Polar Ocean and the Atlantic Ocean. One could also regard the Mediterranean basin as a separate system.

The various objects also contain evident but important properties like depth, length and channel width. Important properties are further the morphology of the river, which can be braided or meandering or can be transformed into a channel by man.

Some of these characteristics can be detected using automated techniques. Radar sensing techniques can reveal very accurately the height of the surface of the earth. The lower temperature of water surfaces with respect to the surrounding land can be detected with high-resolution imagery. Using a combination of these techniques, the expected flow of water from the elevation models and the actual presence can be derived.

Man-made structures like dams, dikes, canals or reservoirs make up another set of components of a water reference. Unlike most natural phenomena, these objects are more difficult to catalogue in an automated manner. For example, canals do not necessarily flow in a specific direction, and the area they drain does not have to be located next to the canal. Finally, there is also a need for an inventory of groundwater resources; however, these very important resources will not be elaborated in this document.

Albeit a very large reference regarding the objects that are to be inventoried, it is expected that only a limited part of the reference needs to be updated on a yearly basis, as influencing surface water flow requires major investments that are not occurring that frequently. Moreover, most possible dam locations in Europe are already exploited, whilst water channels as a carriageway are loosing interest of most investors at the expense of road and train cargo solutions.

Some changes can be expected in, e.g. Spain, where large systems are designed and built to divert fresh water to southern areas in order to extend irrigation schemes and to provide more leisure facilities.

In this overview, the extraction of watersheds and catchments using radar-sensed elevation models will be discussed. Subsequently this will be complemented with techniques which automatically detect water surfaces.

Finally, it will be demonstrated how these results can be used to generate an automated catalogue for detecting water surfaces wide enough to be detected by high-resolution satellites (25 m resolution).

The generation of the catalogue will consider both the so-called Strahler and the Pfafstetter coding mechanisms. These hydrological coding mechanisms allow the information in the catalogue to be grouped in manners that follow hydrological logic. Furthermore, these coding mechanisms can be generated from the topology of the data themselves.

Since the catalogue consists of objects each having a specific geometry, this geometry can also be used to link existing geographic data with the reference and to enrich the reference with such data. Exploitation of these typical GIS functions will be subject to further exploitation of the data. In a sense, both the polygons of the primary catchments and the river segment draining this primary catchment can be used to link in a specific manner ancillary data with the reference.

For example, such relations might yield methods to include waterways or other objects influenced by man to the reference.

In return, existing systems can also be enriched using a geometrical linkage with the catalogue in order to supply them with, e.g., Pfafstetter coding generated using the catalogue.

16.3 RADAR SENSING TECHNIQUES TO DETECT ELEVATION

16.3.1 Introduction

Elevation data were previously generally provided by means of models performing interpolation of measured points in the landscape. Such models are expensive to generate and often resolutions below 250 m are not available. Nowadays radar detection techniques provide an alternative. By detecting the time the transmitted signal returns to the sensor, one can deduct the height of the surface that reflects the signal. Application of radar techniques in spacecrafts give way to create a product of comparable quality throughout the whole area worked through.

However, radar detection detects the height of the surface; hence, the roofs of a group of buildings will be processed as the surface of that particular area. This has to be taken into account when processing radar-derived altitudes instead of the classical method that focuses on points on the soil surface of the earth.

The Shuttle Radar Topography Mission (SRTM) was flown aboard the NASA Space Shuttle Endeavour flight between the 11th and 22nd of February 2000. The mission's Space-borne Image Radar (SIR-C) and dual X-band Synthetic Aperture Radar (X-SAR) were able to collect over 75% of the Earth's surface. The mission covered everything between 60 Degrees North and 56 Degrees South Latitude. The SRTM "finished" data meet the absolute horizontal and vertical accuracies of 20 m (circular error at 90% confidence) and 16 m (linear error at 90% confidence), respectively, as specified for the mission.

The SRTM data were initially delivered by the United States Geological Survey (USGS) in an "unfinished" version. Data for the territory administrated by the United States where post-processed and delivered to the public in a "finished" format.

The European data were delivered in tiles of 1 decimal degree by 1 decimal degree. Each tile contains 1201 rows and 1201 columns. A pixel then represents 1 decimal degree/1200 or 0.000833333 decimal degrees in both directions. This represents around 8000 m^2 in southern Europe and 5000 m^2 in northern Europe.

For the European Window, some surfaces were signalled as "noise" or "voids" (no signal) whilst many water surfaces were displaying a "wave" signal (between −4 and +4 m above sea level).

Therefore, additional processing was needed to interpolate the voids and to "flatten" the water surfaces.

16.3.2 Interpolation of the Voids

The voids mentioned earlier appeared on places where the radar reflection signal was not sufficient to elaborate a clear altitude. In general, this appears on four types of landscapes: areas covered with ice (glaciers), areas with very steep elevations (cliffs and crest lines), flowing water surfaces (rivers) and areas that are too hot (deserts). These areas have in common that the radar signal returned is too diffuse to be interpreted clearly.

To interpolate the pixels marked "void", a linear interpolation was used in which the nearest cells in all directions having valid values where considered and a line toward the nearest valid cell in the opposite direction was determined.

This method, known as linear interpolation, causes a smoothing of the surface. Especially for crest lines, this method can be different from the reality, since linear interpolation will never interpolate to higher elevations than those observed in the surrounding data. For example, using linear interpolation methods the glaciated summits of the Alps will be lower than in reality.

However, for our purpose, to derive the flow direction of water, a linear interpolation was considered as sufficient.

Our method further worked through the interpolation of continued regions of voids.

Starting from the North West row, a region was interpolated and the interpolation results on the edge of the region were taken into account in the next processing iteration. Assuming that voids below 200 m altitude coincide with water surfaces the elevation of the lowest valid point connected to the void region was used as the value for the entire region.

Since the data were delivered in tiles of 1201 rows and 1201 columns (with 1 row and 1 column overlaps), the interpolation was performed by taking 9 neighbouring tiles into computer memory and processing only the centre tile, thus greatly reducing edge issues.

A full elevation model was generated for a window from longitudes 32 West and Latitude 27 North up to Longitude 69 East and Latitude 60 North. To cover Europe up to the North Cape (Northern Norway) readily available elevation models at a resolution of 250 m were acquired for the latitudes between 60 and 72 North. Only for the Northern Russian part of Europe, a 1 km resolution elevation model had to be used.

After the interpolation of the voids, one can proceed to the detection of a direction of flow of any surface within Europe. Using methods described by Soille et al. (2003) rivers can thus be generated from a detailed elevation model.

As mentioned before, the lake, sea and wide (or broad) river surfaces were also detected in order to enhance the elevation model output with additional knowledge from space.

For SRTM generated data, this becomes even more interesting since the data sets follow the reflection of forest crowns and building roofs rather than the surface of the earth. Hence, one can expect less rivers running through towns and forests than is observed in reality.

Figure 16.3 gives an example from Flevoland, which is a reclaimed land (polder) area in the centre of the Netherlands. In the North West area of the image the noise signal, due to the waves of the lake "Markermeer", is visible. The polder itself within the red line contains forest and cities that are marked green, whilst the lower level areas constitute natural reserves and agricultural land and they are coloured blue. Blue colours represent areas below sea level, whilst the green areas are above sea level. The polder however is fully below sea level and has been reclaimed only in 1968. Hence the SRTM signal reveals the forest crowns and buildup area roofs instead of the surface of the earth.

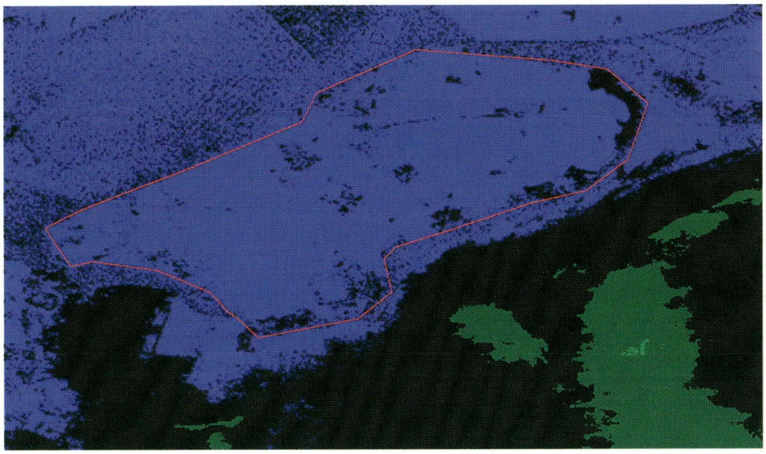

Fig. 16.3. Flevoland, reclaimed land (polder) area in the centre of the Netherlands.

16.4 DETECTION OF WATER SURFACES

16.4.1 Development of a Method to Detect Water

In 1999 the Landsat 7 satellite was launched, the instruments on board of this vehicle were able to shoot imagery of the earth surface with a resolution of 15 and 30 m.

Cloud free imagery of Landsat 7 was acquired for the whole territory of 23 member states of the EU, Bulgaria and Romania.[2] The images were taken in the summers of 1999, 2000 and 2001, normally around 11 o'clock solar time. Each image covers rectangles of around 100 by 100 km of earth surface, and they typically overlap 5 to 10 kilometres with each other. Subsequently these images were rectified and projected to the various national projection systems as commonly used in the member states of the European Union.

The rectified imagery is made available to the public via a web service operated by the European Commission's Joint Research Centre.

The images were made available at a resolution of 25 m and they contain seven bands. The three first bands reflect the human visible light while band 5 contains an infrared reflection between a wavelength of 1.45 and 1.75 micrometers.

After sampling a large number of images, a methodology to detect water surface was developed.

Band 5 yielded, wherever water was present, an apparent clear peak in the lower values. This is illustrated in Figure 16.4, where an image with peak area up to a pixel value of 15 in band 5 of a Landsat 7 image is shown. On the X axis the pixel values are set and on the Y axis, the number of pixels attributed to this value. The data in this graph reflect the pixel distribution of an image taken in June 2000 in Northern Germany. The pixels lower than the lowest value after the first peak (marked in yellow) were extracted as water surface. The rest of the image was interpreted as land (marked green in the graph). The pixel values can however coincide with shadows of small clouds, sometimes still present on the imagery. However, the clouds generate a very high reflection, which can be used as a method to mask the area having cloud cover and the shadow below it. Since the data collection contains geometric overlap for areas with cloud cover, a subsequent image will correct the previous one being partially masked. Since also permanent snow cover yields a cloud reflection, the method could not be applied in areas containing extensive snow cover in the summer months. In Europe, this limitation was observed in the summit area of the Alps, Pyrenees and parts of Northern Sweden.

After having extracted the water surfaces per image, a mosaic was generated per projection system group (see Table 16.2 for the various projection systems used in the 25 member states of the European Union and Romania and Bulgaria).

Normally the generation of a mosaic using various images is hindered by the difference in sunlight and camera angle in a set of images. The method applied solved this limitation by extracting a "low values water signal" per individual image. Hence, the pixels regarded as water were dependent on the total set of characteristics of an individual image. This way specific characteristics like shoot angle and time of the day when the image was

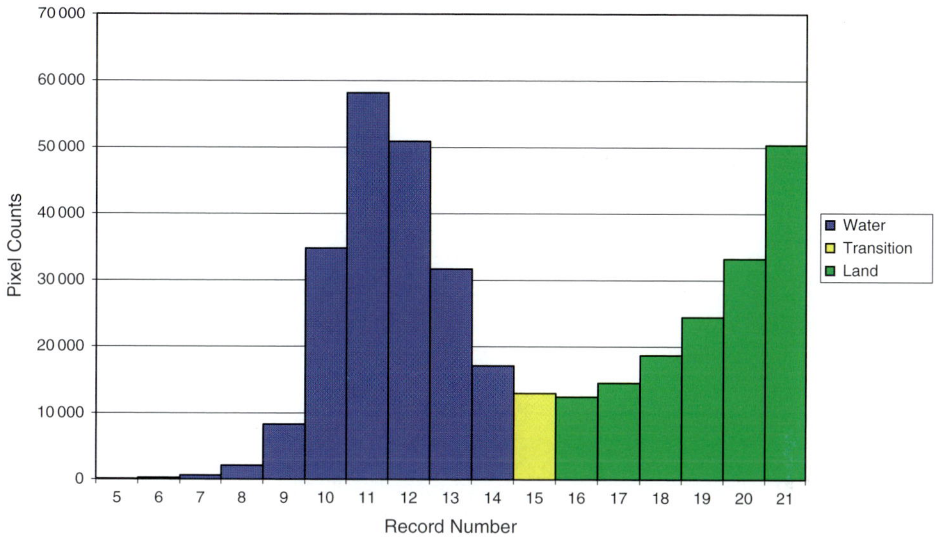

Fig. 16.4. Image with peak area up to a pixel value of 15 in band 5 of a Landsat 7 image.

taken were optimised per individual image and the number of artefacts due to differences in camera angle and daylight fall were acceptable per neighbouring set of images.

A subsequent visual inspection revealed that on latitudes southern than 45 degrees and on some Atlantic coastal areas the differences between water reflection and the neighbouring land was less evident than on northern latitudes. In order to correct for this behaviour, a method was developed that also accepted pixels above the lowest point after the first peak. This method yielded better results but also some shadows of high mountains were detected as water surface. The latter areas were masked out by a simple stratification of the terrain using the SRTM elevation model. The water bodies in the mountain areas on latitudes south of 45 degrees were subsequently detected using the method as laid out in Figure 16.4.

Figure 16.5 shows the detection of land and water in the Burgas area, located in Bulgaria (Black Sea Coast). The dark blue parts contain water detected using Image2000 imagery that was absent in the land water mask used for the processing of CCM version 1. Light green areas were land detected using the Landsat 7 detection but absent in the land water mask previously used. Furthermore, small inland lakes (dark blue) can be observed as new water bodies previously omitted. In addition, piers in the harbour can be automatically detected using the method mentioned before.

16.4.2 Post-processing of the Detected Water Areas

After the detection of the water areas, a mosaic was generated per group of images having the same projection system definition. In order to generate one consistent catalogue these

Table 16.2. Overview of European countries and their specific projection systems, datum's and spheroids as being used in the Image2000 datasets

European Country	Projection System	Datum x,y,z (dx, dy, dz r)	Spheroid	Imagery Source
Albania				UNI
Armenia				—
Austria	Lambert	575.0 93.0 466.0 −5.1 − 1.6 −5.22.5	Bessel	JRC
Azerbaijan				—
Belarus				—
Belgium	Lambert	−113.758268943 59.1656052805 −98.3911345127 −0.5150079386 0.1735134265 −1.7079754953 1.2795	1927	JRC
Bosnia Herzegovina				UNI
Bulgaria	Transverse	WGS 1984	WGS 1984	UNI/JRC
Croatia				UNI
Cyprus				UNI
Czech Rep.	Transverse Mercator	26 − 121 − 78	Krasovsky	JRC
Denmark	UTM	−87 −98 − 121	1927	JRC
Estonia	Lambert	WGS 1984	GRS 1980	JRC
Faroer				None
Finland	Transverse Mercator	−78 −231 −97	WGS 1984	JRC
France	Lambert	−168 − 60 320	Clarke 1884	JRC
Germany	Transverse Mercator	585.7, 87.0, 409.2, −0.524, −0.155, 2.822, 1.0000088	Bessel	JRC
Georgia				—
Greece	Transverse Mercator	−199. 799 74. 281 246. 545 −0.00000497212 0.000000138884 0.000001860967 −0.000000017	Grs80	JRC
Hungary	Hotine Oblique Mercator	52. 684 −71. 194 −13. 975 0.3120 0.1063 0.3729 1.0910	1967	JRC
Iceland				—
Ireland	Transverse	506 − 122 611	Modairy	JRC
Italy	UTM	−87 −98 − 121	1927	JRC
Latvia	Transverse	WGS 1984	GRS 1980	JRC

Table 16.2. (Continued)

European Country	Projection System	Datum x,y,z (dx, dy, dz r)	Spheroid	Imagery Source
Lithuania	Transverse	WGS 84 Seven	GRS 1980	JRC
Luxembourg	Transverse Mercator	−192.986 13.673 −39.309 0.409900 2.933200 −2.688100 0.43000	1927	JRC
Macedonia FYROM				UNI
Malta	Transverse	−87 − 98 − 121	1927	Overlap
Moldova				Overlap
Netherlands	Oblique Stereo-graphic	583.16 26.15 478.54	Bessel	JRC
Norway	UTM	WGS 1984	WGS 1984	UNI
Poland	Transverse	WGS 1984	GRS 1980	JRC
Portugal	Transverse Mercator	504.1 -202.9 563.0	Bessel	JRC
Romania	Transverse	28 −121 −77	Krasovsky	JRC
Russia				UNI
Slovakia	Transverse	26 −121 −78	Krasovky	JRC
Slovenia	Transverse Mercator	668 −205 472		JRC
Sweden	Transverse Mercator	419.384 99.334 591.345 − 0.8504 − 1.8173 7.8622 −0.9950	Bessel	JRC
Switzerland	Hotine Oblique Mercator			Overlap
Turkey				UNI
Ukraine				—
United Kingdom	Transverse Mercator	375 − 111 431	Airy	JRC

groups of images were projected to the SRTM projection system (WGS) and sampled to half the SRTM cell size. This corresponds to 0.00043 decimal degrees, which is approximately 40 m. The overlapping areas were used as samples to test the quality of the geo-referencing of the mosaic.

Figure 16.6 shows a Set of Image2000 Landsat 7 footprints. The area of interest is the Western Coast of Continental Europe. Images delivered in a different projection system are coloured differently. Visible is the area of overlap between the images. This area

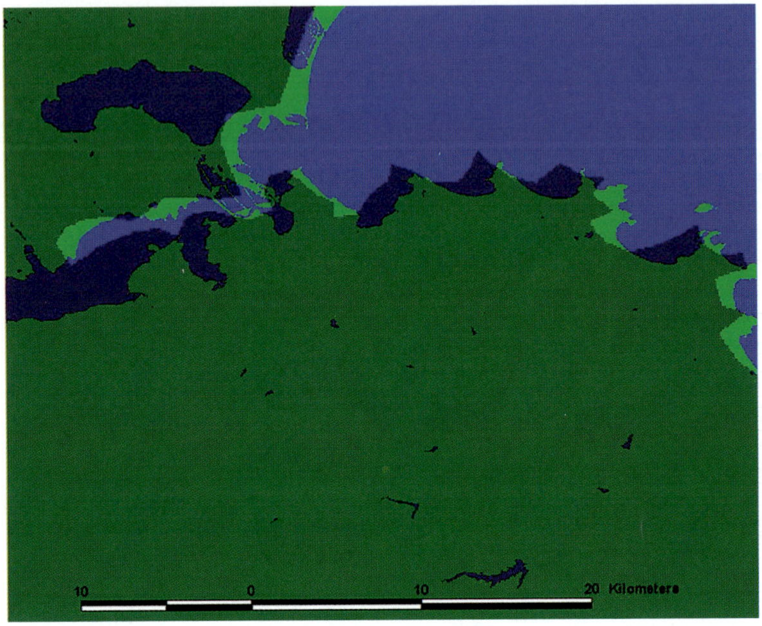

Fig. 16.5. Land and water detection of the Burgas area in Bulgaria (Black Sea Coast).

can be in the order of only 10 kilometres in North–South stretch but half the image (50 km) in East-West direction. The latter overlap is due to the path taken by the satellite vehicle. The path for the next run partially overlaps the previous run. These overlaps are functional for resolving areas with clouds, by taking these from a next image on another moment.

Wherever a low quality was reported, a visual inspection was carried out or additional projection system parameters were assumed.

For various countries, a lack of detail of the provided projection system definition was observed. In general, these omissions concern the definition of the datum of local projection systems. Datum adaptations proved to be necessary for France, Luxembourg, Sweden, Finland, Slovenia, Portugal and Austria. An overview of the various Bursa Wolf parameters of the European Countries is provided in Table 16.2. Note that most countries, due to legacy reasons, have a specific datum, often causing problems in exchange of data since the parameters are omitted and standards regarding projection system definitions are still poorly used by standard software packages. Typically, datum errors account for shifts of dataset in the order of 500 m. Since this distance is twenty times the pixel size of Image2000 data, such a shift was not acceptable for extracting water bodies. If only three parameters are given in the table, then these are to be applied for the so-called Molodenksi transformation. The seven parameters must be applied using a Bursa Wolff transformation. Data taken from the Global Landcover source were all in UTM zones

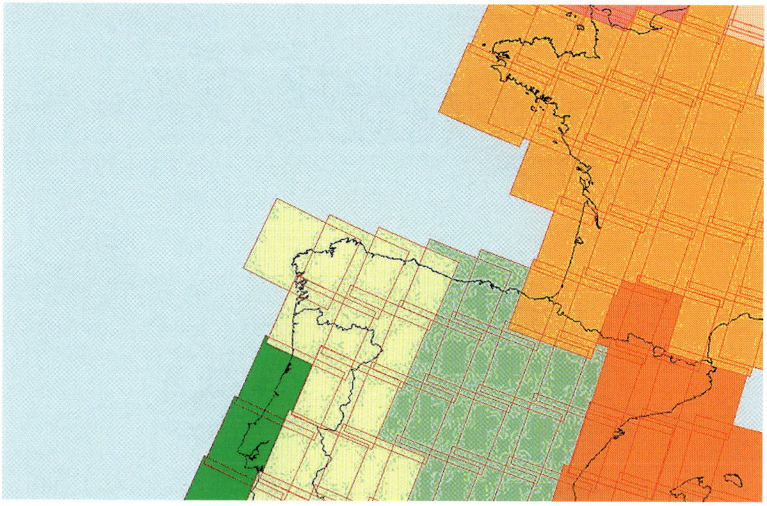

Fig. 16.6. Set of Image2000 Landsat 7 footprints.

on a WGS 1984 datum and spheroid. Note also that a significant number of countries that joined the Union after 2003 apply a standard WGS 1984 datum. The WGS reference system is the bases for most transformation formulas and therefore the datum parameters refer to this arbitrary reference. Thus, where WGS 84 is marked it is implied that the datum transformation is 0,0,0.

The correctly processed mosaics were merged to one European Mosaic collection. Thus, countries that were not participating in the Image2000 data acquisition like Andorra, Malta, Switzerland and Macedonia were also covered. However, large areas of Croatia, Norway, Russia, Turkey, Albania and Serbia and Montenegro were missing since these countries were not fully covered by neighbours involved in the Image2000 data acquisition. Furthermore, no data for Cyprus were available.

In order to complete the catalogue of comparable quality, an additional set of ETM+ detected data were downloaded from the website of the Global Landcover Facility.

Images originating from the same sensor were selected. These were provided in a UTM projection based on the WGS 1984 Datum and Spheroid.

In total over 600 images of 100 by 100 km each were processed per projection system.

The resulting dataset contained thus only one unique typology being "water reflection".

In order to differentiate between sea and lakes, the result was geometrically intersected with an existing catalogue. Every group of pixels that did not touch another group was considered a separate object, called a "feature".

If a feature in the new catalogue intersected with sea or lakes in the existing catalogue, then the whole feature was considered "sea" or "lake".

Linear features like rivers were more difficult to match without error. This was due to the variation in geometrical quality in the existing catalogues.

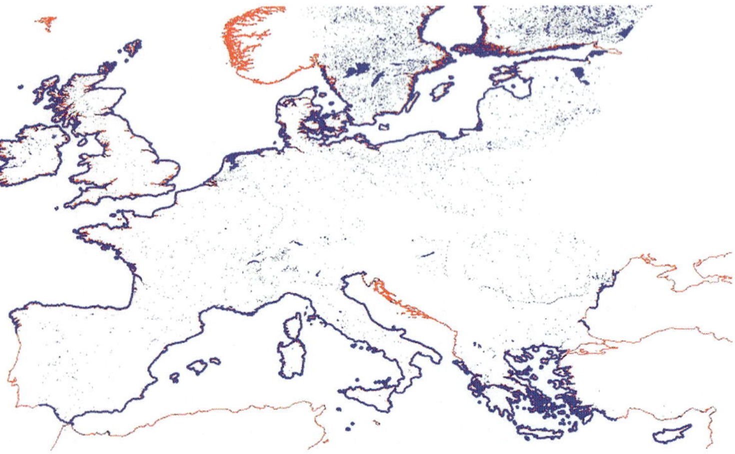

Fig. 16.7. Overview result of the automated water detection of Image2000 Landsat 7 data.

Furthermore, it proved very difficult to delineate the end of a lake or the end of a river mouth. For the latter problem, around 50 river mouths were processed manually. Here a logical (but somewhat arbitrary) division of where the sea stops and where it is regarded as a river was made.

Figure 16.7 shows an overview result of the automated water detection of Image2000 Landsat 7 data. Water surfaces were detected with the method outlined in this chapter and using a buffer of 0.5 decimal degrees around a global coastline. The Portuguese coastline is missing in this version due to projection system definition omission, which was resolved in a later stage. Cyprus and the West Balkan coast were determined using data from the global Landcover facility.

16.5 AUTOMATED CODING METHODS FOR HYDROLOGICAL SYSTEMS

16.5.1 Introduction

The coding of the elements in the catalogue has a dual purpose. It should enhance the automation of data exchange and it should facilitate the maintenance of the features in the system.

A common requirement in assisting the maintenance of the data in an automated system is that the identifiers make sense to the personnel responsible for the updating of the data. Such a human readable code is a code that makes sense to a person who is involved in using the data. A well-known example is the fiscal code made up of a person's birthday and parts of his name.

For hydrological systems, a coding system proposed by Pfafstetter is recommended by the GIS expert group for the WFD. This coding system allows for identification of tributaries with respect to the longest river branch in a river system. The source tributaries are coded 9 while the outlet tributaries are coded 1. The four largest area tributaries flowing to the longest river path are coded 8, 6, 4 and 2. The remainders in between, the so-called interbasins, receive the codes 7, 5 and 3. Using this method, the code can be constructed recursively until every river segment is identified uniquely. Afterwards the position of a segment within the drainage system can be understood without visualisation. Hence, also for monitoring related practices, as labelling water quality samples, such a code can be a basis. Furthermore, small branches that are lacking in a chosen automated representation of a river system can easily be manually coded.

Figure 16.8 gives a representation of the Thames river system with each segment labelled with a unique Pfafstetter code. The hydrological source is identified by 9, while branches near the outlet in the North Sea are coded 1. This map can be generated using the web resource http://wise.jrc.it/data-dist/maps. The background image contains the SRTM elevation model with a colour scheme optimised for the elevation of the Thames basin.

Table 16.3 shows the Thames river system in tabular form. Once the coding mechanism is understood, a user can select a segment relative to a position in the system or can detect an error. Note that the "Thames" started a segment coded 9889, thus the formal riversource does not coincide with the hydrological source in the CCM representation of the riversystem. The hydrological code will be coded 99 and reflect the longest possible flow path in the total system of the Thames.

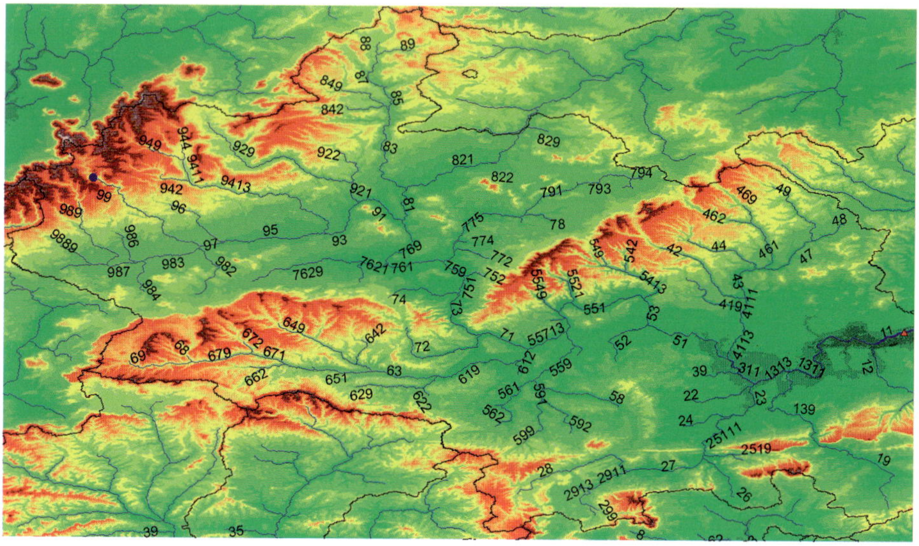

Fig. 16.8. A representation of the Thames river system with each segment labelled with a unique Pfafstetter code.

Table 16.3. Tabular representation of the Thames river system

Name		Inserted on
Thames		2 March 2005
Path from Source to the Outlet		
Pfafstetter	**Sum Length (m)**	**No. of 250 m pixels**
9 889	7 350	1 341
9 881	10 021	445
987	15 658	615
985	16 411	13
983	26 884	1 499
981	27 592	13
97	31 767	252
95	49 533	3 489
93	61 295	1 213
91	71 853	773
769	86 169	1 886
761	96 331	772
759	1 01 411	398
751	1 07 025	495
73	1 20 240	1 451
71	1 29 178	956
55 713	1 37 244	415
55 711	1 41 257	276
555	1 43 067	271
553	1 46 441	200
551	1 60 993	1 429
53	1 68 459	508
51	1 88 160	2 581
311	1 93 447	182
313	1 97 381	256
1 313	2 04 636	1 131
1 311	2 31 378	9 042
11	2 44 063	1 859
1	2 44 188	0
Subsequently flowing into the Southern North Sea with pfafstetter code 5 River with length **2 44 189** m		

16.5.2 *Automated Generation of the Pfafstetter Code*

An additional advantage of the Pfafstetter code is that such a code can be generated automatically provided a consistent drainage network is available for a computer program.

The algorithm was implemented using Oracle Spatial and another version using ArcInfo AML. Such an algorithm requires two geographical datasets. The first one is a set of connected river segments. Secondly, a set of area representation of catchments that are drained by these river segments must be present.

The segments must be related to each other in a network, while the lines should be directed in the direction of flow. The system of segments can be represented as a set of elements linked by from-node and to-node. In this system, one outlet can be found. It is a segment for which to-node a linked from-node cannot be found. All the sources in the system can be found in a similar way. It is the collection of segments having a from-node, for which no segment with a corresponding to-node can be found.

Having these two elements one can determine the longest flow path using common GIS packages, which can determine the length of each segment. The longest flow path starts with the source having the longest flow path to the source. Hence, for all possible sources a flow path to the outlet is to be calculated.

After finding the longest flow path, the four major basins are to be defined. According to Pfafstetter, this is to be determined by the area drained by a tributary on the longest flow path. This area calculation can be done by relating a catchments database to the river segments. This relation can be determined by a geometrical relation between the river segments and the catchments. Typically, the from-node of a river segment will fall in the catchment it drains, while the to-node will link in the next catchment downstream. The area drained by the catchment can consequently be added to the river segment and be propagated to the segments downstream until the longest flow path is reached. The four largest tributaries will then be coded 8, 6, 4 and 2 starting from the source catchments that is coded 9. The tributaries between 8 and 6 will be coded 7, etc. Some rivers like the Danube can yield interesting results with regard to the geographical particularities. In the Belgrade area (Serbia and Montenegro), three major tributaries of the Danube join the main stream. In this case, the Pfafstetter code system has problems in handling the coding automatically. This is due to the geological and geographical situation when the water leaves the Pannonian Plains. All major tributaries river join here the mainstream in order to leave a plain that used to be an inner sea in the Pliocene. This is illustrated in Figure 16.9. The Danube (coded 7115) enters from the North West and receives on the border with Croatia a major tributary from the West, the Drave, (purple spot) after only 10 segments the Tisza coded 4 enters the Danube (green spot). This tributary is coded 4117. Subsequently, a tributary from the South West, the Sava arrives at the main stream (red spot) leaving only two segments for the interbasins coded 36 and 35. Every tributary downstream of Beograd will be coded with a 1 and further like, e.g. the Tamis.

Table 16.4 shows the distribution of the coding in the Danube, Rhone and Rhine system. The Pfafstetter system subdivides large tributaries along the longest flow path. The repartition of these tributaries depends on geological phenomena that prevented the system from evolving to a mature river system.

16.5.3 Non-flowing Water

A drawback is that the Pfafstetter logic can be applied only for a river system in a status of natural flow. Diverging points in deltas cannot be coded and the many man-made

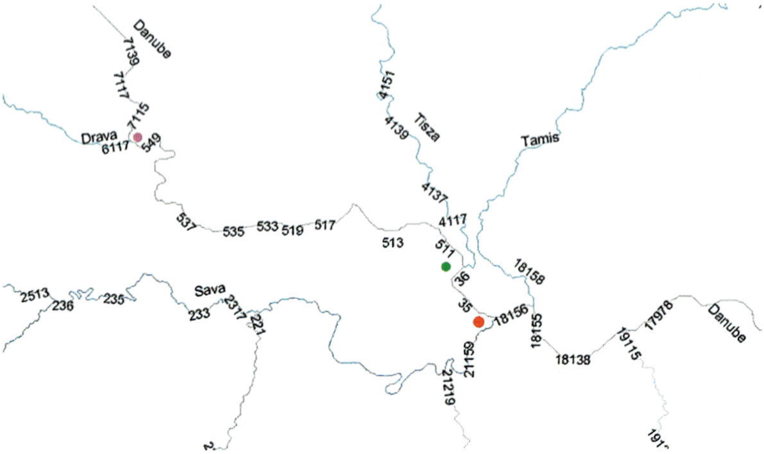

Fig. 16.9. The Beograd area.

Table 16.4. Distribution of the coding in the Danube, Rhone and Rhine system

Pfafstetter code	Danube segments	Rhone segments	Rhine segments
1 Outlet	9584	138	598
2 Tributary	6451	2613	893
3 Interbasins	5	1191	456
4 Triburary	5379	3878	872
5 Interbasins	37	304	67
6 Tributary	4912	3462	492
7 Interbasins	5892	293	956
8 Triburary	5524	341	2686
9 Source	2245	292	2594

adaptations of the natural drainage system cannot be accommodated. Furthermore the Pfafstetter methodology cannot be applied systematically for other relevant objects in the water repository like lakes, swamps, tidal areas and groundwater bodies.

16.5.4 Coding of the Water Systems

In order to extend the Pfafstetter methodology, a coding mechanism based on a similar concept to code the major water systems in Europe has been developed.

A water system is then defined as a set of river branches having one outlet into the sea.

The seas around Europe were coded 1 to 9, using a subdivision of the seas based on marine treaties. In order to remain consistent with the Pfafstetter methodology, the Black

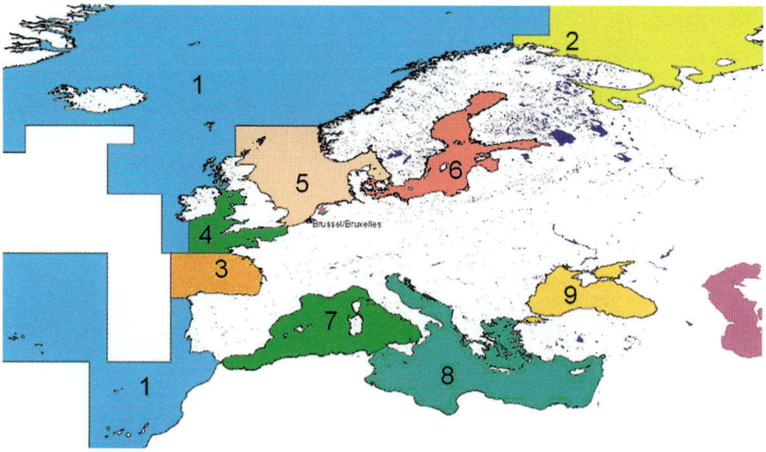

Fig. 16.10. Sea number according to Pfafstetter logic in the seas around Europe.

Sea was coded 9 and the Atlantic Ocean coded 1. Within every sea, the four largest sea outlet water systems are selected based on their area size.

These four sea outlets were coded 2, 4, 6 and 8 clockwise (2 in the North East, 4 in the South East, etc). The smaller sea outlet systems in between are then coded 1, 3, 5, 7 and 9. This is shown in Figure 16.10. The seas further from the ocean have a higher number that the ocean coded 1 (outlet). Around Europe the following seas are numbered 9 Black Sea, 8 Mediterranean East, 7 Mediterranean West, 6 Baltic Sea, 5 North Sea, 4 Irish Sea, 3 Golf of Biscay, 2 White Sea and 1 Atlantic Ocean. The Caspian Sea and other areas not draining at the surface to the ocean could be numbered 0. The delineations of this map are based on fishery treaties. One could also argue to base such contours on hydrologically meaningful boundaries, e.g. based on main currents in the seas.

Figure 16.11 displays the land subdivision of the European sea outlets following the sea subdivision explained in the previous map. Using the code 9 for all systems draining to, for example, the Black Sea allows for direct identification of who is draining finally where. The small pink polygons are areas without surface flow to the sea. Notorious are the Burgenland Sea, Lake Trasimeno and Lake Prespa. In our example, these are coded 0. Furthermore, it is interesting to note that only the Ebro, Po, Tiber, Rhone and Maritsa are considerable rivers flowing into the Mediterranean Basin. European waters flow mainly into the landlocked East and Black seas.

Such a coding can be partially automated, if based on a consistent computer representation of the area drained by a connected river system. It yields at all levels direct insight into how the river system is related to the sea it is flowing into and which other systems influence the same sea. Finally, it can form a basis to code also the components of the water repository which are not flowing as consistent as a river to the sea.

Figure 16.12 shows a second-level coding of the sea-outlets. In this map the four largest systems per sea are coded 2, 4, 6 and 8 and the remaining ones in between 1, 3, 5,

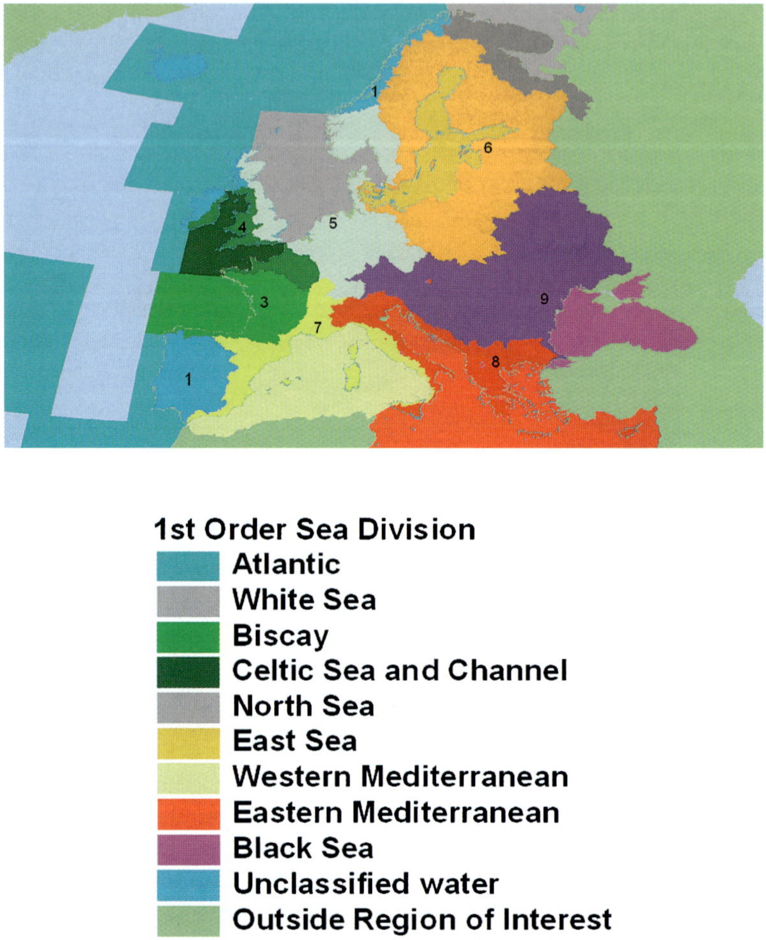

Fig. 16.11. The land subdivision of the European seaoutlets following the sea subdivision explained in Figure 16.10.

7 and 9. System 2 is the most North East system, 4 the most South East, etc. The small systems between system 2 and 4 are coded 3. Thus, the Rhone receives number 4 and the area between the Rhone and the Ebro, number 5. The Ebro, being one of the four largest systems draining into the Western Mediterranean (code 7), is thus assigned the full code 76. This method is not fully automatic due to the islands which are to be coded arbitrarily.

Table 16.5 gives a list of second-level Pfafstetter attribution for sea outlet systems. The Strahler order represented in the last column is an index counting the number of tributaries on the main path, hence an index for the complexity of the hydrological system.

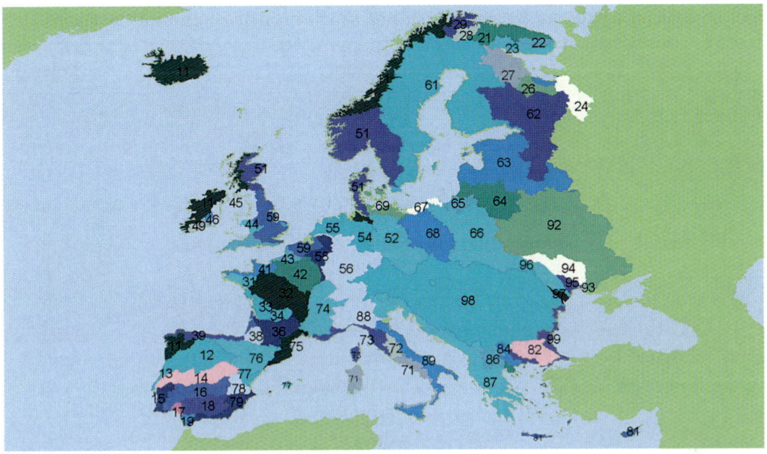

Fig. 16.12. This map shows a second level coding of the seaoutlets.

A combination of this Pfafstetter code with the coding at river segment level allows for a systematic unique identification for any water body in Europe. For example, the source of the Rhone would be 7469.

16.6 INCLUSION OF OBJECTS INFLUENCED BY MAN

Europe contains various impressive alterations of the natural landscape. In general, these regard mainly waterworks. However, also extensive build-up areas like the Greater London Area, the Paris Metropolitan area and the Ruhr area city chain yield poor results for automatic detection of water surfaces. The Radar Elevation detection will detect a group of high building blocks as a mountain summit. Rivers might be flowing through tunnels under the town and not be detectable anymore for satellite sensors.

Furthermore, also landscapes along the North Sea coast can be altered so extensively that water is not flowing downward anymore, but pumped upward. In these polders, hydrological coding does not make sense and detection rules are difficult to impose on a model. The reverse alteration exists as well: In Northern Italy extensive areas are put under water during the growth season in order to allow for rice cultivation.

The polder areas can be below sea level as well, in addition also drained floodplains along rivers can function as polders.

With notion to the sea level, it is also important to note that the Caspian Basin is 15 m below the level of the Black Sea and as such the most extended area below sea level of the planet.

For a system that needs to support also environmental policies, these areas are all of utmost importance. It is often here, in the lowest areas, that an effect of mismanagement with regard to water purity and/or supply is felt first and most dramatically.

Table 16.5. List of second level Pfafstetter attribution for Seaoutlet systems

Pfafstetter code	Name	Strahler order
12 (Atlantic Ocean)	Duero/Douro	7
14	Tajo/Tejo	6
16	Guadiana	6
18	Guajalquivir	6
22 (White Sea)	Ponoy	4
24	Onega	5
26	Vyg	5
28	Patsoyoki	5
32 (Bay of Biscay)	Loire	6
34	Dordogne	5
36	Garonne	7
38	Adour	6
42 (Celtic Sea, Channel)	Seine	6
44	Severn	5
46	Barrow	4
48	Suir	4
52 (North Sea)	Elbe	6
54	Weser	5
56	Rhine	7
58	Meuse	5
62 (East Sea)	Neva	7
64	Nemunas	5
66	Wisla	6
68	Oder	6
72 (Western Mediterr.)	Tiber	6
74	Rhone	8
76	Ebro	7
78	Jugar	5
82 (Eastern Mediterr.)	Maritsa/Evros	6
84	Strimonas	6
86	Axios/Vardar	6
88	Po	7
92 (Black Sea)	Dnepr	7
94	Pivdennyy Buh	5
96	Dnister	5
98	Danube	8

Figure 16.13 shows the rebuilding of a dike after inundations in Zeeland (South West of the Netherlands) in 1953. This area of the Netherlands has been dramatically influenced by man-made alterations to the natural water flow. Notice that delta areas can also change quickly due to natural phenomena, like the change of a channel after sedimentation of sands.

Fig. 16.13. Rebuilding a dike after inundations in Zeeland (South West of the Netherlands) in 1953.

16.7 DISPLAY OF RESULTS

An important tool in understanding the impact of environmental policies is through maps. The WFD anticipates on this by obliging member states to deliver map data in digital format.

Using a common geometric reference allows to display such data in a comparable way for various regions, despite the difference in quality of the underlying data.

By using the lowest level or Strahler order 1 catchments as an integrative building block, one can map hydrologically relevant information on a shared base.

The following example displays landscape diversity according to the so-called Shannon index. Such an index can help to understand where vulnerable headwaters exist.

The index is calculated by adding for each patch type present in the proportion of area covered, multiplied by that proportion expressed in natural logarithm, according to the formula:

$$SHDI = -\sum_{i=1}^{m}(Pi * \ln Pi)$$

m = number of patch types

P_i = proportion of area covered by patch type (land cover class) i

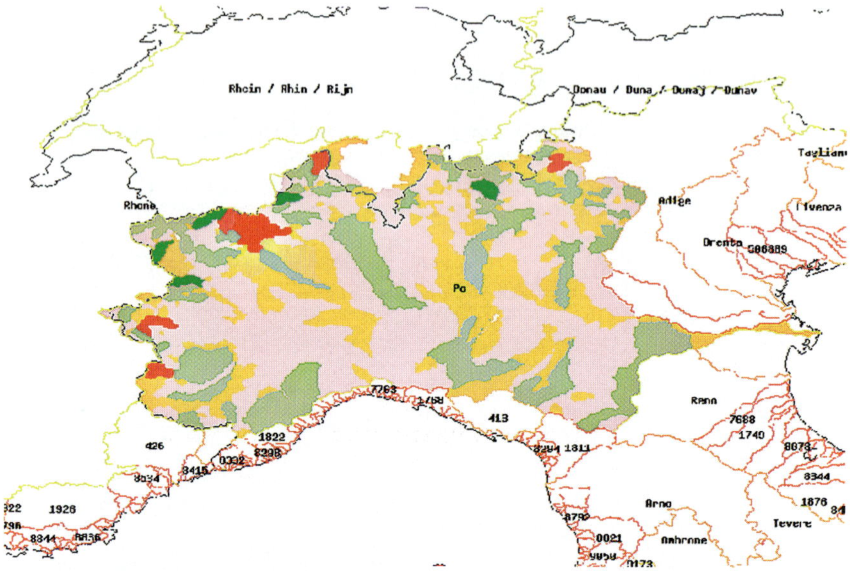

Fig. 16.14. The Shannon landscape diversity index of the Po watershed system in Northern Italy.

Figure 16.14 shows the Shannon landscape diversity index of the Po watershed system in Northern Italy. This map is based on Corine Landcover vector data of 1990. The red valleys (Gressoney) in the North West have the highest landscape variability. The polygons are reported on a Strahler level 3 repartition of the Po system. In CCM 1 the Po system was subdivided up to 7 Strahler orders. This index is based on the patch density of Landcover data.

Another example is based on the Pfafstetter coding mechanism, and this mechanism, as explained earlier, displays the four largest systems within a large system along the longest path. Hence, an immediate insight is gained into the hydrologic importance of the various regions with a sea outlet system. Figure 16.15 shows the Rhone Pfafstetter subdivision at the highest level using CCM version 1. The nine systems can be easily recognised and show that the outlet system (red) is small in this particular case. The hydrological source of the system coded 9 does not coincide with the upper Rhone valley (coded 6, the river named Rhone is the dark blue line) but with the Saone valley. The subdivision of these nine systems can be used for additional breaking up of, e.g., river management plans, since it makes sense to consider each system as an independent system up to the main flow path.

16.8 CONCLUSIONS AND RECOMMENDATIONS FOR FURTHER RESEARCH

In this chapter it was demonstrated that the main water resources of Europe could be extracted using space born data. There are several advantages and disadvantages to a

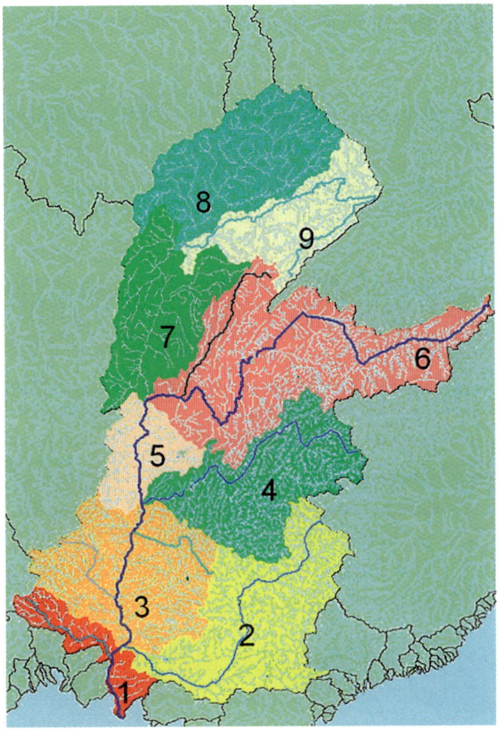

Fig. 16.15. Rhone Pfafstetter subdivision at the highest level using CCM version 1.

build-up of a catalogue using space born data sources in comparison with vectorized map sets. In Table 16.6, a list of those advantages is represented. The table is not exhaustive, but is meant to display the different potentials of the various options.

Since both space born and vectorized data sources coexist and will continue to coexist, it is recommended to use the coordinates of given objects to be linked to river segments and catchments in an automated manner to the space born catalogues. In order to assist this automation the Space born River segments and Catchments database needs to become immutable and coded in a concise manner. To make the lowest level catchment areas immutable, specifically for the artificially drained areas, additional research is needed.

Linkage of datasets should not be limited to geometrical intersection only. For linear objects like rivers and canals, also names, lengths, sources and outlets can give means for linkage. Polygon objects as lakes and coastal areas can best be intersected using classical methods in which the geometry of the most detailed source can be preserved, eventually one can quantify the geometrical linkage in percentages of area covered, thus allowing for additional means to understand the quality of the linkage.

Table 16.6. Advantages and limitations of space born data sources of water catalogues

Property	Space born water detection	Vectorized datasets
European coverage	Up to 60 degrees	Varied quality
Link between catchments and rivers	Guaranteed but not valid in karstic and artificial managed areas	Catchments delineations often not existing
Level of detail	Maximum 90 m, perhaps in future to 30 m	Reference to various map scales, not seamless
Updateable	Clear reference to image date	Variable per map sheet, often not digital incorporated
Availability	Almost free (only processing time and algorithm implementation to be counted for)	Depending on market potential, in general river data in dry or remote areas are free, in urban areas they are not
Topological intelligence	Can be derived from the elevation model, for example flow direction	Has to be brought in mainly manually. In some member states this is already done
Metadata	Easy to determine once the source of the image is known and the processing methods are automated	Difficult to determine due to need to access many natural languages and different systems filled by operators originating from different mapping traditions

16.9 GLOSSARY OF TERMS

Karstic Land where water occasionally flows below the surface due to natural processes.

Intersection Computer operation in which the geometrical representations of two objects is cross-referenced.

Landsat 7 Vehicle containing sensors with up to 12 m resolution detection of the earth surface.

Masking Not processing a marked area on a geo-referenced image.

Pit Area of land not draining on the surface to the sea, in Europe Lake Trasimeno, Burgenland See and the Caspian Sea are examples of pits. Also polders can be modelled as Pits.

Polder Reclaimed land, out of which rainwater has to be pumped into an overflow, polders are often below sea level.

Region Continued area of pixels in an image having the same pixel value.

River segment Part over a river up to the next tributary.

Sea outlet Area of land draining to one point where the water enters the sea. Such a system is an abstraction of reality, since it omits the diversion of waters, which is common in a delta.

WEBSITES

http://image2000.jrc.it website containing an interface to register a data request for image2000 data.

http://edc.usgs.gov/products/elevation/srtmbil.html website containing description of the SRTM product.

http://glcfapp.umiacs.umd.edu:8080/esdi/index.jsp Earth Science Data Interface (ESDI) at the Global Land Cover Facility.

http://europa.eu.int/comm/agriculture/publi/landscape/index.htm From Landcover to Landscape Diversity in the European Union.

http://wise.jrc.cec.eu.int Water Information System for Europe Upload site.

NOTES

[1] The Water Frame Work directive is signed by the 25 member states of the European Union, Bulgaria, Norway, Romania and Switzerland.

[2] Of the 25 EU member states, Malta and Cyprus did not participate in the project mentioned.

Chapter 17

PERSPECTIVES FOR INFORMATION SYSTEMS FOR WATER

P. Haastrup

17.1 INTRODUCTION

In the EU there is currently a strong concerted effort to improve the information flow and accessibility of water data, and in fact EU water policies have evolved fast in the last few years. The Water Framework Directive (also known as the WFD or Directive 2000/60/EC) (EU Parliament & Council, 2000) is a very important piece of legislation for the protection and improvement of the quality of all water resources within the European Union. Considering the complexity of the technical issues, it is recognised that a common approach to some of these challenges would allow the sharing of expertise, information, experience and resources whilst facilitating, as far as possible, a coherent and harmonious implementation of the Directive. This of course has implications for the information systems.

According to DG ENV (2005), the WFD is complemented by earlier directives which will continue to cover some aspects of the water environment (which will actually act as instruments linked to measures under the WFD): The Urban Waste Water Treatment Directive, the Bathing Water Directive, the Drinking Water Directive and the Nitrate Directive. New policies are being developed for the protection of the oceans and seas bringing together the current patchwork of policies, legislation, programmes and action plans at national, regional, EU and global level for the marine environment. In addition, an initiative has been taken to consider the feasibility of a new EU legislation related to sustainable flood prevention and control. Even outside the Union, the European Union Water Initiative (EUWI) is aiming to bring together EU institutions and Member States, developing country government agencies, civil society, financial institutions, as well as the EU's water industry to raise the priority given to water and sanitation issues in the context of poverty reduction efforts in partner countries. Other European policies (i.e. Common Agricultural Policy, Thematic Soil Strategy, etc.) have in many instances a role to play in the protection of water resources and can contribute or reinforce the achievement of the objectives of the water policies.

In view of boosting information, exchanges of knowledge and good practices, and to summarise key information related to water policies on a regular basis, a wider initiative to modernise the collection and dissemination of information on water policy across

Europe – the Water Information System for Europe (WISE), is being constructed. It should be noted that at the moment the WISE system is considered an umbrella for a number of existing initiatives, and is intended to structure and streamline the processes.

Member States, the European Commission (in particular DG Environment, the Joint Research Centre and Eurostat) and the European Environment Agency (EEA) have agreed that there is a need for streamlining the reporting process, gathering more useful and relevant information and making the exchange process as efficient as possible using modern technology for the information on water.

17.2 THE EDEN-IW PROJECT IN THIS CONTEXT

When the Environmental Data Exchange Network for Inland Water (EDEN-IW) project was conceived in early 1999, the Water Framework Directive had not yet been adopted.

The Directive was on its way through the legislative system, and at that time it was not yet possible to foresee neither the adoption date, nor the development of the details of the Directive. Now, six years later the picture that emerges is that the Water Framework Directive has shaped the regulatory landscape in the European Union, to such an extend that it is the single most important piece of legislation in the water area. Daughter directives are on the way, old legislation is expected to be incorporated or removed, and the decision-making processes linked to the Water Framework Directive are in place, with a Common Implementation Strategy for all Member States.

At the same time, the WFD requires a significant scientific and technical input in the process, in order to arrive at the goals of the directive, which simplified can be stated as "good ecological status" for Europe's water bodies by 2027.

This is where the results from the EDEN-IW project, which you now have in your hand, are valuable, both for the positive and the negative experience gained, in order to facilitate the future work.

17.3 THE INFORMATION SYSTEM

While the EDEN-IW project was exploring the boundaries of scientific and technological knowledge to share information from different databases, the regulatory processes in the WFD also moved to a concept of shared information.

In 2004, the Water Information System for Europe was born, and the ideas and concepts from the EDEN-IW experiences were parts of that process.

The key to the functionality of the WISE system is significantly different from the EDEN-IW project. It is mostly linked to the need for the WISE system to stay with proven technologies that can be maintained for a significant period of time, and that has shown themselves to be robust and reliable.

The WISE system will contain various categories of data, similar to those handled by the EDEN-IW project, that is:

1. Text
2. Data (in form of more or less complex data bases) and
3. Geographical information (vector data, shape files, images, etc.).

While a series of principles for the complete handling of the WISE information can be set up, differences exists between the categories.

In practical terms, the text and data categories are relatively dense, and can easily be moved across the network and thus easily duplicated for safety, independence or re-elaboration purposes, while large geographical or image data sets should not be transferred in the same way, but rather used as a service on demand, according to the INSPIRE principles.

The WISE information system now contains a series of integrated geographical data (see Figure 17.1 with the data on main rivers). As can be seen from the figure, the density of rivers reported to the system is not yet uniform, with significant differences in the granularity level between Member States. It is of course the intention that the flow of data from the Member States to the EU level should be more or less uniform, in order

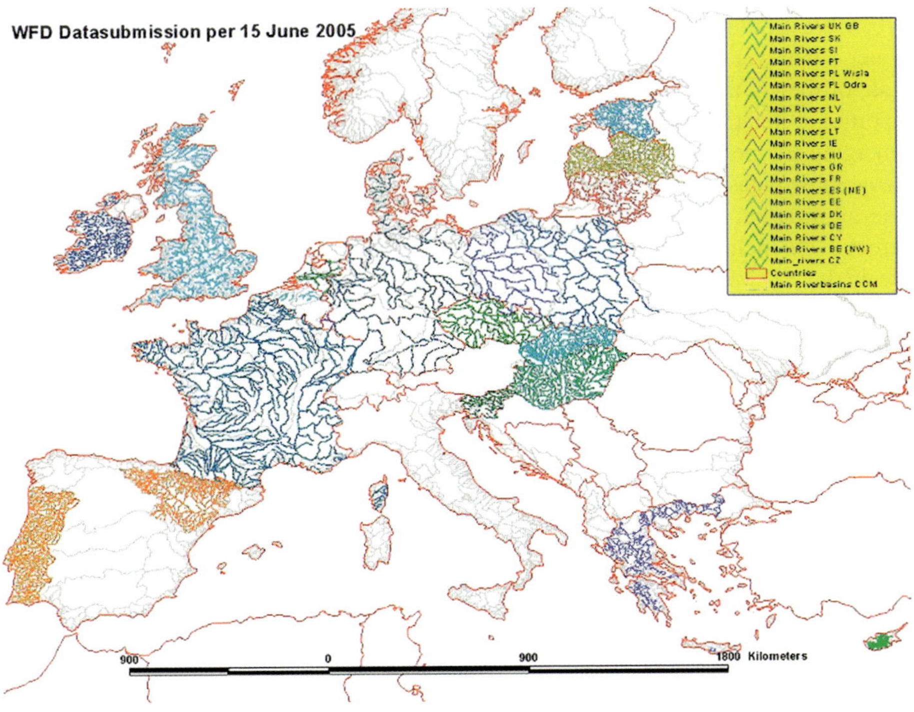

Fig. 17.1. Main rivers in Europe as hosted by the Water Information System for Europe on June 15, 2005.

to minimise the reporting burden on the Member States and allow for easily harmonised analysis to be carried out covering all EU territory.

Currently the data flowing into the WISE systems are related to the 2004 and 2005 submission of basic data, while the 2006 reporting is linked to river basins and measurement activities. From 2007, environmental quality of water data will also be provided.

17.4 THE FUTURE DEVELOPMENT OF WISE

In the future, it is the intention to develop WISE into a shared information system where all water-related data on European level are linked with each other, where data flow is harmonised and streamlined and where a variety of users can access the information for different purposes.

WISE will cover all water related information stemming from EU water policy and it should be extended to cover other European water-related datasets. By geo-referencing most of these data, WISE will become an important building block for the INSPIRE.

The future WISE is a modular system, which can be extended and adapted to emerging needs, but it would include, at least, data and information related to:

- Compliance reporting under all EU water legislation (Water Framework Directive, Urban Waste Water Treatment, Nitrates Directive, Bathing Water Directive, Drinking Water Directive, etc.).
- State-of-the-environment and trend information collected in addition to compliance reporting obligations and currently incorporated in EIONET Water.
- Emerging EU policies such as the Marine Strategy and the Flood Action Programme.
- Geographic datasets linked to hydrographical aspects in the context of INSPIRE.

The information should be gathered in a coordinated way and be made available to different users including the general public. The planned timetable for WISE is given in Table 17.1.

The further development of WISE requires various tasks and activities to be implemented.

For the coming years, the following, non-exhaustive, list of tasks will need to be addressed, in particular:

- Development of the interface for access to WISE and launching and maintaining the public service

Table 17.1. Planned WISE development

2005	Development of an implementation plan for WISE 2006–2010 including the distribution of roles of the different partners
2006	Open public access to selected information in WISE
2006–2010	Extension of WISE to cover all water policies and all water-related dataflows
2007	Joining WFD and EIONET dataflow
2010	WISE fully operational and sole instrument to submit compliance reports by Member States

- Integration of various other existing databases and dataflows into WISE
- Development of indicators based on dataflow including development of specific indicators linked to thematic areas, e.g. water environment and health or urban water management
- Thematic and systematic assessments of compliance
- Visualisation of data for public
- Analysis and modelling using geo-referenced data, including scenarios.

The WISE information system will be developed in steps. The precise steps are still under discussion. However a number of principles and guidelines are clear, also based on the experiences with the EDEN-IW project.

First of all, the information stored should be readable both for humans and for computers.

Secondly the information should also be readable in 50 years time, with subsequent conditions for the encoding.

In addition a high degree of independence among the nodes should be achieved, especially for text and numeric (tabular) data. For geographic data, the INSPIRE principle of not copying unnecessarily large data source should be followed.

Part of this detailing of requirements will include the main protocol rules. This allows existing systems to become "WISE compatible" and new systems to be developed "WISE compatible".

One approach to detailing the specifications could be to subdivide the communication based on usage, e.g. "meta data publishing", "uploading" and "data access".

Uploading concerns the action when data in a specific format is sent to the WISE system and is subsequently checked against predefined business rules.

Meta data publishing is a functionality in which the WISE system allows specific users to understand which data exist in the specific component of the WISE system and if these data are available to this user.

Subsequently the exchangeable data can be accessed by a system with such ability. This should be predefined beforehand; however, how and when specific data are harvested depends on the data access system and its specific purpose.

The subdivision of WISE in these three core components does not only reflect the organisational requirement that a number of mandated organisations might host a specific upload process of WISE data. It will also allow for system specialisation along the lines of uploading systems with a focus on transaction speed and security, publishing systems with a focus on reliability, and harvesting systems which will play a data–warehouse and public access role with regard of the WISE data elements and others.

Therefore it is to be expected that WISE nodes will be developed focussing on different functions of an information system such as input or output functions.

Automated data access functions should then allow the various systems to be synchronised and be perceived by the end user as a single computer system.

In addition to the synchronisation approach explained above, there is a need to allow for a higher degree of data modelling or harmonisation by integrating datasets physically, carefully considering their methodology and applying necessary analysis techniques.

Also a looser approach of coupling data providers by integrating commonly referenced data will become a function of a comprehensive WISE system.

17.4.1 Software Architecture

A software architecture that enables the fulfilment of the requirements should be developed taking the existing software architecture into account. The plan for the preparation phase should include the completion of this activity.

The implementation of the WFD and the reporting on the State of the Environment requires that a number of geographically dispersed organisations being part of public administrations or mandated by administrations will have to upload and/or exchange water related data.

These organisations will typically have diverse software and hardware requirements and very specific priorities regarding how information systems are to be implemented.

The WISE system architecture resolves a set of specific needs that emerge from this complex business model in which WISE will function.

Furthermore WISE has to resolve solutions for data uploading, data access and exchanging of future directives.

Therefore the WISE system architecture anticipates a system architecture in which a *data exchange protocol* plays a central role.

17.4.2 Implementation and Test of WISE Synchronisation in an Iterative Approach

The preparation phase shall be subdivided into iterations leading eventually to the WISE system as of late 2006.

Two specific components are to be developed and subsequently realised. The first component consists of a set of specifications allowing the sharing of essential processing data between WISE nodes. Subsequently this component can be extended to exchange the data between the various WISE nodes.

Secondly a WISE interface design is to be developed, allowing any user to interact with a WISE compatible system in a consistent way.

Once these two base components have been developed, the multitude of components solving the business issues of WISE can be developed and slotted in the total system. Figures 17.2, 17.3 and 17.4 give a possible outcome of the various WISE modules that make up the final system by 2010. It shows that every module can contain a large set of independent structures but that essential processing information is exchanged and that, if allowed, data can be exchanged by one WISE node and harvested by another. Furthermore the current situation lacks output modules allowing mapping and extended data mining.

In Figure 17.2 the current situation of the WFD prototype, serving for the input server is shown. In parallel the input system EIONET/WATER, serving the needs of EEA, is

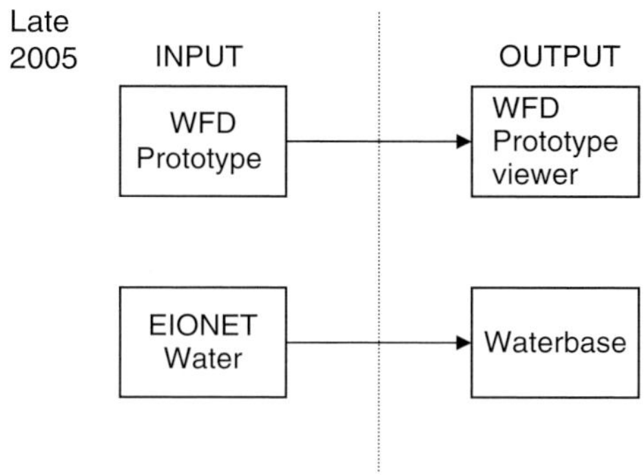

Fig. 17.2. Current situation, showing two non-linked services.

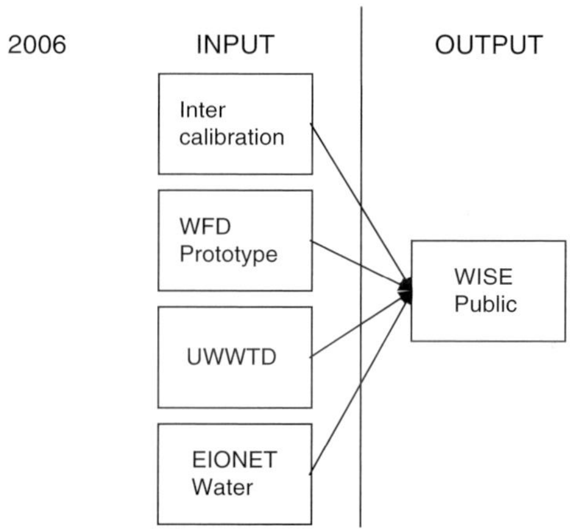

Fig. 17.3. Expected water data content by the end of 2006 – showing one public access points to several distributed services.

shown. At the current moment in time, the data delivered by Member States to the WFD prototype is under development.

During 2005 and 2006 the main development foreseen is the construction of an output server, here called "WISE Public", which must show the feasibility of integrating the input from various servers into one public service. This is shown in Figure 17.3, with potential input from various information sources: The WFD prototype, EIONET/Water, the Intercalibration servers and the Urban Waste Water Treatment Directive servers.

In Figure 17.4 the vision of the situation in 2010 is shown, with additional input systems being integrated and with additional possibilities for extracting and analysing the output, e.g. state of the environment and research purposes.

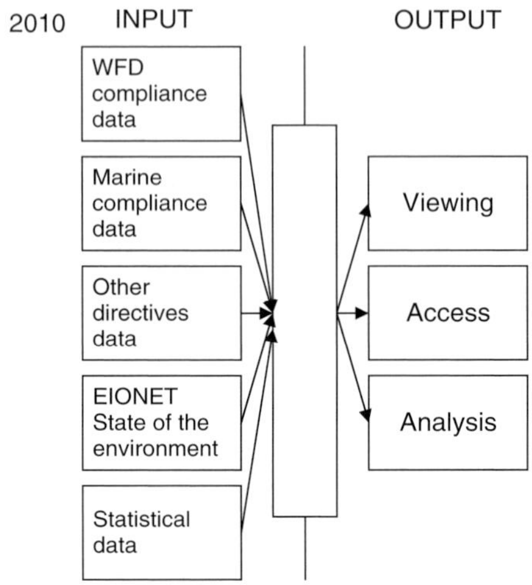

Fig. 17.4. Projected situation around 2010, showing the option to apply different, distributed functions on several input services.

17.4.3 WISE System Data Exchange

In order to exchange mostly tabular (non spatial) data between independent data (input) nodes, an approach inspired by the open software community has been adopted. This approach implies that a WISE node (either for input or indeed for output, for instance through a public website) maintains its own copy of the data (mostly non-spatial) being available from various specialised data providers called "WISE node". Along with this an exchange/synchronisation mechanism will ensure that the data stored in the public website are at all times in synchronisation/harmonisation with the data being available from the other WISE nodes. The synchronisation mechanism must be provided in such a way that it will enable the introduction of additional data providers into it.

Each WISE node decides in principle itself what data it chooses to expose/publish through the synchronisation mechanism, and also which parts of the exchanged/synchronised data pool it chooses to keep a (synchronised) copy of at its own location/node. This is shown in Figure 17.5.

In this view there is no single master of all data. Therefore, the harmonisation of the data will, in principle, be determined by the level of semantic detail and formality of the exchange protocol.

It is the responsibility of each WISE node to maintain a translation mechanism between these exchanged and synchronised data and a consolidated data model (and database) stored at the local facility/server.

Two essential functions enable the existence of this WISE data provision mode. These are the basic exchange protocols and their supportive shared data models and the interface guidelines. Once the basic exchange protocol is designed, the existing WISE components

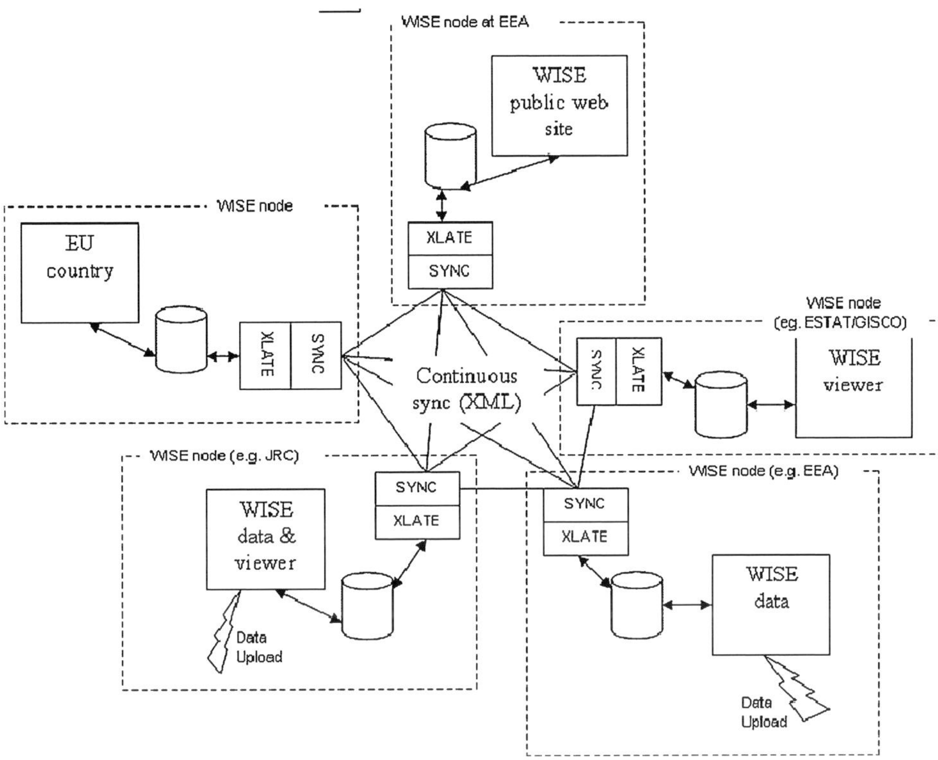

Fig. 17.5. Data exchange between independent data nodes, of mostly non spatial data, through synchronisation of independent data providers. The published information will be available at all WISE nodes, and can be incorporated in the local system as required (WISE, 2005).

need to be adapted to this protocol using the new interface guidelines. For new WISE nodes the protocols will be guidelines for development of the new node in such a way that the system becomes WISE compatible.

The data model should also include a definition of required metadata.

The basic data, of for instance a concentration value, could be tagged in XML to assure consistency. For the data set as a whole, a metadata description is needed. This metadata set could be based on a "Dublin Core" set modified appropriately. The metadata set will need to be made by organisms holding the data to be exchanged, and encoded in XML to a common standard.

An exchange method is to be specified allowing these elements to be exchanged online between various systems. Table contents will be exchanged between various computers connected to each other only using a TCP/IP protocol. Regarding the exchange, security issues must be addressed and real-time exchange, e.g. password updating, is to be

DG ENV, 2005. WISE Newsletter. "The Bulletin of the Water Information System for Europe" Vol. 1 – no. 1, June (P. Quevauviller, Ed.) European Commission, DG Environment, Unit D2. http://ec.europa.eu/environment/water/pdf/wise_june_2005.pdf.

EDEN-IW, 2002. Multilingual Glossary of Inland Waters, D15. http://www.eden-iw.org.

EDEN-IW, 2003. Structured List Integration into IW – Distributed Semantic Hybrid Agents application – D17. http://www.eden-iw.org.

EDEN-IW, 2004. Project Home Page: http://www.eden-iw.org. Available from August 2004.

EEA, 1999. European Environment Agency (Felluga, B. and Batschi, W.-D., Eds) *GEMET, General European Multilingual Environmental Thesaurus*. ixpp. + Volume 1: *Systematic List of Descriptors*, 44pp.; Volume 2: *Thematic List of Descriptors*, 78pp.; Volume 3: *Alphabetical List of Terms*, 550pp.; Volume 4: *Concordance List*, 127pp.; Volume 5: *Multilingual List of Descriptors*, 536pp. CNR, UBA – Umweltbundesamt, Version 2.0. EEA, Copenhagen, August.

EEA, 2001. Global implementation plan (GIP) for the IDA Reportnet project. Copenhagen. http://eea.eionet.eu.int:8980/Public/irc/eionet-circle/eionet-telematics/library?l=/reportnet_process/idasreportnetsgipstacdoc/_EN_1.0_&a=I

EEA, 2002. EEA support to the European Community in reporting obligations within the framework of international environmental conventions – Legislative instruments, international programmes and conventions. Copenhagen. http://reports.eea.eu.int/Technical_report_No_62/en

EEA, 2004. Streamlining of reporting: A first level analysis ROD Analysis. Version 1.0 draft. Copenhagen. http://eea.eionet.eu.int/Members/irc/eionet-circle/nfp-eionet/library?l=/nfp_eionet_group/2005_february/ session_3/streamlining_reporting/_EN_1.0_&a=i

EEA, 2005. Environmental dataflows under European reporting obligations. Version draft 1d. European Environment Agency internal paper, Copenhagen.

EEA's Multilingual Glossary, 2004. http://glossary.eea.europa.eu/EEAGlossary/

EEA-TF, 1995a. European Environment Agency – Task Force (Felluga, B., Ed.). *MET – Multilingual Environmental Thesaurus*. Classification Scheme. CNR, Roma, June, pp. 3 + 90.

EEA-TF, 1995b. European Environment Agency – Task Force (de Lavieter, L., Ed.; Lucke, S., Plini, P., and Felluga, B., Editors For the Italian equivalents). *MET – Multilingual Environmental Thesaurus*. Part 1, English; Part 2, Français; Part 3, Deutsch; Part 4, Nederlands; Part 5, Italiano; Part 6, Norsk; Part 7, Dansk; Part 8, Español. NBOI, Nederlandse Bureau voor Onderzoek Informatie / EEA-TF – European Environment Agency – Task Force, Brussels; European Environment Agency, Copenhagen. Amsterdam, pp. (English) vi + A-78; B-112; C-56; D-199, Total 445, Grand total, pp. ca. 4 000. November.

EEA WB, 2004. Waterbase. http://glossary.eea.eu.int/EEAGlossary/W/Waterbase

EU Cordis, 2004. Project Database. http://dbs.cordis.lu/search/en/simple/EN_PROJ_simple.html. Available August 2004.

EU Leonardo da Vinci Project, 2004. Aqualingua [Multilingual dictionary of technical words and expressions related to the water sector]. http://scripts.oieau.fr/aqualingua/pres/index.htm

EU Parliament & Council, 2000. Directive 2000/60/EC of the European Parliament and of the Council establishing a framework for the Community action in the field of water policy (In short, the EU Water Framework Directive [or even shorter the WFD]).

EU Parliament & Council, 2003. Directive 2003/4/ EC of the European Parliament and of the Council of 28 January 2003 on public access to environmental information.

EU Parliament & Council, 2004. Proposal for a directive of the European Parliament and of the Council establishing an infrastructure for spatial information in the Community of 23 July 2004 (In short, the INSPIREdirective).

European Parliament and the Council, 2000. Directive 2000/60/EC of the European Parliament and the Council of 23rd October 2000 establishing a framework for Community action in the field of water policy. *Official Journal of the European Communities*, 22 December, 72pp.

EU-WFD, 2004. Water Resources Framework Directive – Priority List. http://www.oieau.fr/ciedd/contributions/at1/contribution/eea.htm

EU WFD, 2005. http://europa.eu.int/comm/environment/water/water-framework/overview.html

Felluga, B. and Plini, P., 2004a. EDEN-Inland Waters TRS, Terminology Reference System. pp. 1–3504. CNR – Consiglio Nazionale delle Ricerche, IIA – Istituto sull'Inquinamento Atmosferico, UTA – Unità Terminologia Ambientale, Rome, 2004. http://www.eden-iw.org/contents/publication.html.

Felluga, B. and Plini, P., 2004b. EDEN-IW, Environmental Data Exchange Network – Inland Waters Extra Document E17. Quadrilingual List of Determinands and Complementary Parameters (English, Danish, French, Italian). Version 2004-09-09, 373pp. CNR – Consiglio Nazionale delle Ricerche, IIA – Istituto sull'Inquinamento Atmosferico, UTA – Unità Terminologia Ambientale, Rome, 2004. http://www.eden-iw.org/contents/publication.html.

Felluga, B. and Plini, P., 2005. EDEN-IW Terminology Reference System. Reference Language: English. Complete equivalence in Danish, French and Italian. The Environmental Data Exchange Network for Inland Water (EDEN-IW) Project of the Information Societies Technology (IST) Programme. Deliverable E27.

Felluga, B., Plini, P., Lucke, S., and Palmera, M., Eds, 1975. EARTh, Environmental Applications Reference Thesaurus© Pp. vi + Volume 1: *Elenco Sistematico*, 37pp.; Volume 2: *Elenco tematico*, 159pp.; Volume 3: *Elenchi Alfabetici brevi, Italiano-Inglese e Inglese-Italiano*, pp. 113 + 113; Volume 4: *Elenco alfabetico bilingue*, 1418pp.; Volume 5: *Elenco delle Concordanze*, 115pp.; Volume 6: *Elenco Termini liberi – assegnati a liste – esclusi/cancellati*, 20pp.; Total: vi + 1975. CNR, Roma, Aprile 2002 (A GEMET-revised edition, 5300 terms). http://www.t-reks.cnr.it.

Felluga, B., Palmera, M., and Lucke, S., 1989. SBD – Sistema Bilingue di Descrittori per l'indicizzazione, la categorizzazione e la codificazione dei termini ambientali / BDS – Bilingual Descriptor System for Indexing, Categorizing and Codifying Environmental Terms. Roma, CNR-ITBM, pp. xxiii + 278, 1989.

Felluga, B., Gauthier, T., Gunesh, A., Haastrup, P., Neophytou, C., Plini, P., Poslad, S., Preux, D., Santouridis, I., Stjernholm, M., and Würtz, J., 2003. "Environmental Data Exchange for Inland Waters using Independent Software Agents. Scientific Results in the period July 2001 to June 2002". EUR 20549EN.

Felluga, B., Lucke, S., Palmera, M., Plini, P., de Lavieter, L., and Deschamps, J., Eds, 1994. Thesaurus per l'ambiente – Versione quadrilingue / Thesaurus for the Environment – Quadrilingual Version/Milieu-thesaurus – Viertalige vertaling / Thesaurus für die Umwelt. CNR-SIAM & CNR-UPIS. CD-ROM Edition, Milan, 1994. (Includes the contents of: Felluga, B., de Lavieter, L., Deschamps, J., Lucke, S., and Palmera, M., Eds. Edizione pilota, Vol. 1/3, pp. i–xiv + 700; Pilot Edition, Vol. 2/3, pp. xv–xxviii + 684; Proefuitgave, Vol. 3/3, pp. xxix–xxxiv+672, Total pp. i+xiv+xxviii+xxxiv + 2056, 1991, Roma, CNR-ITBM.)

Fensel D. and Brodie M.L., 2003. *Ontologies*. ISBN: 3540003029. Springer-Verlag, Berlin, Germany.

Ferber, J., 1999. *Multi-Agent Systems*. Addison-Wesley, Harlow, UK, 1999.

Finin, T., Labrou, Y., Mayfield, J., 1997. KQML as an Agent Communication Language. In *Software Agents*, Bradshaw, J.M. (Ed.), MIT Press, Menlo Park, CA, USA, pp. 291–316.

FIPA IEEE Computer Society standards organization, 2002a. FIPA ACL Message Representation in XML Specification. Retrieved July 29, 2005, from http://www.fipa.org/specs/fipa00071/

FIPA IEEE Computer Society standards organization, 2002b. FIPA Agent Message Transport Envelope Representation in XML Specification. Retrieved July 29, 2005, from http://www.fipa.org/ specs/fipa00085/

FIPA IEEE Computer Society standards organization, 2002c. FIPA Subscribe Interaction Protocol Specification. Retrieved July 29, 2005, from http://www.fipa.org/specs/fipa00035/

FIPA IEEE Computer Society standards organization, 2005a. Foundation for Intelligent Physical Agents standards activity Home Page. http://standards.computer.org/fipa/. Accessed 2005

FIPA IEEE Computer Society standards organization, 2005b. The Foundation for Intelligent Physical Agents. Retrieved July 29, 2005, from http://www.fipa.org

FIPA-OS, 2005. FIPA-OS Features, http://fipa-os.sourceforge.net/features.htm.

FIPA001, 2005. FIPA Abstract Architecture Specification No. FIPA00001. Accessed from http://standards.computer.org/fipa/.

FIPA023, 2005. FIPA Agent Management Specification No. FIPA00023. Accessed from http://standards.computer.org/fipa/.

FIPA067, 2005. FIPA Agent Message Transport Service Specification No. FIPA00067. Accessed from http://standards.computer.org/fipa/.

FIPA071, 2005. FIPA ACL Message Representation in XML Specification No. FIPA00071 Accessed from http://standards.computer.org/fipa/.

FIPA085, 2005. FIPA Agent Message Transport Envelope Representation in XML Specification No. FIPA00085. Accessed from http://standards.computer.org/fipa/.

Foster, I., Kesselman, C., and Tuecke, S., 2001. The Anatomy of the Grid: Enabling Scalable. Virtual Organizations. *International Journal of High Performance Computing Applications*, Vol. 15(3), pp. 200–222. www.globus.org/research/papers/anatomy.pdf

French Parliament, 1992. Law 92-3 of 3 Jan 1992 on water.

Galbiati, L., Zaldivar, J.M., Somma, F., Bouraoui, F., Moren-Abat, M.C., Bidoglio, G., and D'Eugenio, J., 2005. Pilot River Basin Outcome Report, Testing of the WFD Guidance Documents. Report EUR 21518 EN. 107pp.

GDAL, 2005. GDAL OpenSource Community. GDAL – Geospatial Data Abstraction Library Homepage. Retrieved July 29, 2005, from http://www.gdal.org

Gruber, T.R., 1993. A Translation Approach to Portable Ontologies. *Knowledge Acquisition*, Vol. 5(2), pp. 199–220.

Haastrup, P. and Würtz, J., 2001. "Pilot Implementation of Agent Technology for Environmental Data Exchange Network for Inland Water", EUR 19919EN.

Haastrup, P. and Würtz, J., 2002. "Environmental Management of Water Data in EU Using Agent Technology". Paper presented at "Environmental Communication in the Information Society". September 25–27, 2002, University of Technology, Vienna, pp. 70–77 (Part 1).

Iglesias, C.A., Garijo, M., Gonzalez, J.C., 1998. A Survey of Agent Oriented Methodologies. In *Proceedings of the 5th International Workshop on Intelligent Agents: Agent Theories, Architectures, and Languages (ATAL-98)*, pp. 317–330.

Immon W.H., 1996. *Building the Data Warehouse*, 2nd edition, New York: Wiley.

Infoterra's EnVoc thesaurus, 1997. UNEP, United Nations Environment Programme – Infoterra. EnVoc – Multilingual Thesaurus of Environmental Terms. UNEP, Nairobi, May 1997, pp. xix +248. ISBN 92 0807 12608.

ISO/TC 211 Geographic information/Geomatics, 2005. ISO/TC 211 Geographic information/ Geomatics Homepage. Retrieved July 29, 2005, from http://www.isotc211.org

JADE, 2005. JADE Open-source Community. JADE (Java Agent DEvelopment Framework) Retrieved July 29, 2005, from http://jade.tilab.com

JATLite, 2005. Java agent Template light, JATLite home page. http://java.stanford.edu/java_agent.html

Java Servlet-SUN, 2005. Sun Microsystems. Java Servlet Technology. Retrieved July 29, 2005, from http://java.sun.com/products/servlet/index.html

JAVA-RMI-SUN, 2005. Sun Microsystems. Java™ Remote Method Invocation (JAVA-RMI). Retrieved July 29, 2005, from http://java.sun.com/j2se/1.4.2/docs/guide/rmi/spec/rmiTOC.html

JDBC-SUN, 2005. Sun Microsystems. JDBC API Tutorial and Reference – Introduction. Retrieved July 29, 2005, from http://java.sun.com/docs/books/jdbc/intro.html

JENA, 2005. HP Labs Semantic Web Research. Jena Semantic Web Toolkit. Retrieved July 22, 2005, from http://www.hpl.hp.com/semweb/jena.htm

Jensen, S., Saarenmaa, H., and Martin, J., 2002. Information Infrastructure for European Environmental Reporting – The Contribution through Reportnet – Keynote for Enviroinfo Vienna 2002. In *Environmental Communication in the Information Society*, Pillmann et al. (Eds), Published by IGU/ISEP: International Society for Environmental Protection, Vienna, pp. 29–38.

Joint US/EU ad hoc AML Committee, 2001. The DAML+OIL extension to RDF Schema. Retrieved March, 2001, from http://www.daml.org/2001/03/daml+oil-index

JSSE-Sun, 2005. Sun Microsystems. Java Secure Sockets Extensions (JSSE). Retrieved July 29, 2005, from http://java.sun.com/j2se/1.4/docs/guide/security/jsse/JSSERefGuide.html

Kashyap, V. and Sheth, A.P., 1994. Semantics-based Information Brokering. In *Proceedings of 3rd International Conference on Information and Knowledge Management (CIKM)*, November.

Kerschberg, L., Gomaa, H., Menascé, D., and Yoon, J.P., 1996. Data and Information Architectures for Large-Scale Distributed Data Intensive Information Systems. In *Proceedings Eighth International Conference on Statistical and Scientific Database Management*, June 18–20, Stockholm, SE.

Klein, M. 2001. XML, RDF, and Relatives. In *Intelligent Systems, IEEE*, Vol. 16(2), March/April, pp. 26–28.

Klein, M., Kiryakov, A., Ognyanov, D., and Fensel, D., 2002. Ontology Versioning and Change Detection on the Web. In *Proceedings of the 13th European Conference on Knowledge Engineering and Management, EKAW-2002*, Springer, LNAI, Madrid, Spain.

Labrou, Y., Finin, T., and Peng, Y., 1999. The Current Landscape of Agent Communication Languages. *Intelligent Systems, IEEE Computer Society*, Vol. 14(2), March–April, pp. 45–52.

Lalement, R. and Lagarde, P., 2005. "architecture du système d'information sur l'eau, livre vert (Architecture of the Water Information System, Green Book)" Ministère de l'Ecologie et du Développement durable, MEDD/DE/SIE 2005-1, January, 36pp.

LAWA, 2001. LAWA: Working Group of the German Federal States on Water Problems. Guide to the Implementation of the EC Water Framework Directive. Part 3: Preparation and Guidelines for Drawing Up a Management Plan. Länderarbeitsgemeinschaft Wasser (LAWA) (Working Group of the German Federal States on Water Problems) – Subcommittee of the EU Liaison Committee on preparations for the technical and legal implementation of the EC Water Framework Directive, February 20, 70pp.

MapServer, 2005. MapServer OpenSource Community. MapServer OpenSource development environment Homepage. Retrieved July 29, 2005, from http://mapserver.gis.umn.edu

Martin, D.L., Cheyer, A.J., and Moran, D.B., 1998. Building Distributed Software Systems with the Open Agent Architecture. In *Proceedings of the Practical Application of Intelligent Agents and Multi-Agent Systems 1998*, Nwana, H.S. and Ndumu, D.T. (Eds), London, UK, 355–376.

McBride, B., 2002. Jena: A Semantic Web Toolkit, *IEEE Internet Computing*, Vol. 6(6), pp. 55–59.

McIlraith, S.A., Son, T.C., Zeng, H., 2001. *EEE Intelligent Systems*, Vol. 16(2), pp. 46–53.

Multilingual Descriptor System (MDS) of the European Commission, 1983. CEC – Commission of the European Communities. MDS, Multilingual Descriptor System (MDS) for the European Inventories on the Environment. Peter Peregrinus Ltd. Publ./Commission of the European Communities, Unwin, Old Woking, Surrey, England, Pilot Edition, EUR 8638 EN, pp. iv + (DA-v + 38; + NE-v + 38; + EN-iv + 44; + FR-v + 43; + DE-v + 34; + IT-v + 44; + MDS 212) (total: iv + 453).

Nodine, M.H., Unruh, A., 1997. Facilitating Open Communication in Agent Systems: The InfoSleuth Infrastructure. In *Proceedings of the 4th International Workshop on Intelligent Agents IV, Agent Theories, Architectures, and Languages,* Singh, N., Roa, A., and Wooldridge, M. (Eds), pp. 281–295.

Nodine, M.H., Fowler, J., and Perry, B., 1999. Active Information Gathering in InfoSleuth. In *2nd International Symposium on Cooperative Database Systems for Advanced Applications CODAS*, Wollongong, Australia, pp. 15–26.

Norup, B. (Ed.), 1996. NERI, National Environmental Research Institute. Guidelines for data collection for the Dobris+3 Report. Final Draft. NERI, Copenhagen, pp. 186, September 1996.

Noy, N.F. and Klein, M.C.A., 2004. Ontology Evolution: Not the Same as Schema Evolution. *Knowl. Inf. Syst.* Vol. 6(4), pp. 428–440.

Noy, N.F. and McGuinness, D.L. 2001. *Ontology Development 101: A Guide to Creating Your First Ontology*. Retrieved July 29, 2005, from http://protege.stanford.edu/ publications/ontology_development/ontology101.html

Odell, J., Parunak, H.V-D., and Bauer, B., 2001. Representing Agent Interaction Protocols in UML. In *Agent-Oriented Software Engineering*, Ciancarini, P. and Wooldridge, M. (Eds), Springer-Verlag, Berlin, *Proceedings of 22nd International Conference on Software Engineering*, pp. 121–140.

OGP, 2005. OGP Surveying and Positioning Committee Homepage. [Note: OGP was formed in 2005 by the absorption into OGP of the now-defunct European Petroleum Survey Group (EPSG).] Retrieved July 29, 2005, from http://www.epsg.org

OGR, 2005. OGR OpenSource Community. ISO/TC 211 Geographic information/Geomatics Homepage. Retrieved July 29, 2005, from http://www.remotesensing.org/ogr/

OntoWeb-D1.4, 2005. OntoWeb Project Deliverable 1.4, A Survey on Methodologies for Developing, Maintaining, Evaluating and Re-engineering Ontologies. Available from http://www.ontoweb.org.

OntoWeb-D2.2, 2005. OntoWeb Project Deliverable 2.2, Successful Scenarios for Ontology-based Applications V1.0. Available from http://www.ontoweb.org.

OntoWeb-D4, 2005. OntoWeb Project Deliverable 4.0, Ontology Language Standardisation Efforts. Available from http://www.ontoweb.org.

Open Geospatial Consortium, 2005. Open Geospatial Consortium Homepage. Retrieved July 29, 2005, from http://www.opengeospatial.org

OWL-W3C, 2002. W3C, OWL Web Ontology Language 1.0 Reference, W3C Working Draft, 29 July 2002. Retrieved from http://www.w3.org/TR/2002/WD-owl-ref-20020729/

Passin, Thomas B., 2004. *Explorer's Guide to the Semantic Web*. Manning Publications Co, New York, NY, USA, ISBN 1932394206.

Petersen, T. (Ed.), 1994. *AAT – Art and Architecture Thesaurus*. Oxford University Press, New York, Vol. 1, pp. xxix + 455; Vol. 2, pp. 533; vol. 3, pp. 586; Vol. 4, pp. 586; Vol. 5, pp. 546.

Plini, P. and Felluga, B., 2000. T-REKS, Thesaurus-based Reference Environmental Knowledge System. TDCNet, Conférence pour une infrastructure terminologique en Europe. 13–15 Mars 2000, Maison de l'UNESCO, Paris, France. http://www.unilat.org/ dtil/etis/actasTDCnet/feluga.htm.

Plini, P. and Felluga, B. (Eds), 2003. Ministero dell'Ambiente e della Tutela del Territorio – CNR. STeRNA – Sistema Terminologico di Riferimento Nazionale per l'Ambiente. Ministero dell'Ambiente e della Tutela del Territorio – CNR, Versione 31 Maggio 2003. http://sterna.iia.cnr.it/

Plini, P., Lucke, S., Baffioni, C., and Felluga, B., 2001. T-REKS: A Contribution to the Environmental Information Management through a Computer-Supported Modular Knowledge Organisation System for the Environment. *15th International Symposium Informatics for Environmental Protection*, Zurich, October 9–12, *Sustainability in the Information Society*. Hilty, L.M. and Gilgen, P.W. (Eds), pp. 691–698. http://www.empa.ch/deutsch/fachber/empasg/iep01.

Plini, P., Lucke, S., Palmera, M., and Felluga, B. (Eds), 2006. *EARTh, Environmental Applications Reference Thesaurus*. CNR, Consiglio Nazionale delle Ricerche, Rome, 2006 (8000 terms; in preparation). http://uta.iia.cnr.it/earth-eng.htm.

Poslad, S., 2002. Developing FIPA-based Agent Applications. In *4th EASSS*, European Agent Systems Summer School, July 8–12, Bologna, IT.

Poslad, S. and Charlton, P., 2001. Standardizing Agent Interoperability: The FIPA Approach. In Michael Luck, Vladimír Maŕík, Olga Stepánková, and Robert Trappl (Eds), Multi-Agent Systems and Applications, 9th ECCAI Advanced Course, ACAI 2001 and Agent Link's 3rd European Agent Systems Summer School, EASSS 2001, Prague, Czech Republic, July 2–13, 2001, Selected Tutorial Papers. Lecture Notes in Computer Science 2086 Springer 2001, ISBN 3-540-42312-5, pp. 98–117.

Poslad, S. and Willmott, S., 2003. Modeling Agent Services for Open Environments, Parts 1 & 2. *5th EASSS*, European Agent Systems Summer School Tutorial, Barcelona, Spain, February 2003.

Poslad, S., Buckle, P., and Hadingham, R.G., 2000. Open Source, Standards and Scaleable Agencies. *Int. Conf. Autonomous Agents*, Workshop on Infrastructure for Scalable Multi-agent Systems, Vol. 1887, pp. 296–303.

Protégé, 2005. Protégé Ontology Tool Retrieved July 22, 2005, from http://protege.stanford.edu/

Quevauviller, P., Balabanis, P., Fragakis, C., Weydert, M., Oliver, M., Kaschl, A., Arnold, G., Kroll, A., Galbiati, L., Zaldivar, J.M., and Bidoglio, G., 2005.

Science-Policy Integration Needs in Support of the Implementation of the EU Water Framework Directive. Research & Technology Integration in Support of the European Union Water Framework Directive. *Environmental Science & Policy*, Vol. 8(3), pp. 203–326.

Raskin, J., 2000. *The Humane Interface: New Directions for Designing Interactive Systems*. 1st edition, Addison-Wesley Pub. Co., Boston, MA, USA, ISBN: 0201379376.

RDF-W3C, 2005. Resource Description Framework (RDF). Retrieved July 22, 2005, from http://www.w3.org/RDF/

Reddy, M.P., Prasad, B.E., Reddy, P.G., and Gupta, A., 1994. A Methodology for Integration of Heterogeneous Databases. In *Knowledge and Data Engineering, IEEE Transactions*, Vol. 6(6), December.

RMI Proxy – Telekenesis, 2005. Telekenesis. RMI Proxy Homepage Retrieved July 29, 2005, from http://www.rmiproxy.com

SANDRE, 2005. SANDRE Service National d'Administration des Données et Referentiels de l'Eau. Retrieved July 22, 2005, from http://www.sandre.eaufrance.fr

Sardon, A.M. (Ed.), 1995 Ministère de l'environnement. Lexique environnement – Planète. Tome 1, *Liste alphabétique*, pp. 83; Tome 2, *Liste thématique*, pp. iv + 186. Ministère de l'environnement, Paris, Décembre 1995.

Searle, J.R. 1969. *Speech Acts*. Cambridge University Press, 1969.

Smith, M.K, McGuiness, D., Volz, R., and Welty, C., 2002. Web Ontology Language (OWL) – Guide Version 1.0. Available from http://www.w3.org/TR/2002/WD-owl-guide-20021104/.

ACRONYMS

ACC	Agent Communication Channel
ACV	Agent Communication Visualizer
ANSI	American National Standards Institute
API	Application Programming Interface
CNR	National Research Council, Italy
CORBA	Common Object Request Broker Architecture
DA	Directory Agent
DAML	DARPA Agent Mark-up Language
DARPA	Defence Advanced Research Projects Agency, US
DB	Database
DEM	Digital Elevation Model
DF	Directory Facilitator
DG-INFSO	Directorate General Information Society, European Commission
DG-ENV	Directorate General Environment, European Commission
DMZ	Demilitarised Zone (in network)
DOM	Document Object Model
DTD	Document Type Definition
DTO	Data Transfer Object
EARTh	Environmental Applications Reference Thesaurus
ED	European Dynamics, Greece
EDEN	Environmental Data Exchange Network
EDEN-IW	Environmental Data Exchange Network for Inland Water
EEA	European Environment Agency, Denmark
EGV	EDEN Global View
EIONET	European Environment Information and Observation Network
ETC/Water	European Topic Centre for Water
EU	European Union
FIPA	Foundation for Intelligent Physical Agents
FTP	File Transfer Protocol
GEPA	Gregory E. Pitts Associated, US
GEMET	General Multilingual Environmental Thesaurus
GIS	Geographic Information System
GML	Geography Mark-up Language
HCI	Human Computer Interaction
HTA	Hierarchical Task Analysis

HTML	Hyper Text Mark-up Language
HTTP	Hyper Text Transfer Protocol
ICMP	Internet Control Message Protocol
IEC	International Electrotechnical Commission
IES	Institute for Environment and Sustainability, JRC Italy
IIOP	Internet Inter-ORB Protocol
IOW	International Office for Water, France
ICT	Information and Communication Technologies
INSPIRE	Infrastructure for Spatial Information in Europe
ISO	International Organisation for Standardisation
IST	Information Society Technologies
JADE	Java Agent Development Framework
JDBC	Java Database Connectivity
JRC	Joint Research Centre Ispra, European Commission, Italy
KB	Knowledge Base
KOS	Knowledge Organisation System
KQML	Knowledge Query Manipulation Language
KR	Knowledge Representation
LBNL	Lawrence Berkeley National Laboratory, US
LDV	Local Database View
LRU	Least Recent Used
MAS	Multi-Agent Systems
MCC	The Microelectronics and Computer Technology Corporation, US
MDS	Multilingual Descriptor System
NERI	National Environmental Research Institute, Denmark
OGC	Open Geospatial Consortium
OIL	Ontology Inference Layer
OLM	Ontology Lifecycle Management
ORB	Object Request Broker
OWL	Ontology Web Language
PNG	Portable Network Graphics
QMUL	Queen Mary University of London, UK
RA	Resource Agent
RDF	Resource Description Framework
RDBMS	Relational Database Management System
REST	Representational State Transfer
RMI	Remote Method Invocation
RNDE	National Network for Water Data, France
setenv	setenv, L.L.C., US
SIE	National System for Information on Water, France
SMTP	Simple Mail Transfer Protocol
SOAP	Simple Object Access Protocol
SQL	Structured Query Language

SSL	Secure Sockets Layer
SSOA	Semantics Service Oriented Architecture
SRTM	Shuttle Radar Topography Mission
TA	Task Agent
TCP/IP	Transmission Control Protocol/ Internet Protocol
T-REKS	Thesaurus based Reference Environmental Knowledge System
TRS	Terminology Reference System
UA	User Agent
UDP/IP	User Datagram Protocol/Internet Protocal
UML	Unified Modeling Language
UNEP	United Nations Environmental Protection
URI	Universal Resource Indicator
URL	Universal Resource Locator
US-DOD	Department of Defence, US
US-EPA	Environmental Protection Agency, US
USGS	United States Geological Survey
UTM	Universal Transverse Mercator
WGS	World Global System
WFD	Water Framework Directive
WFS	Web Feature Service
WISE	Water Information System for Europe
WMS	Web Map Service
WSDL	Web Services Description Language
XMDR	Extended Metadata Registry
XML	Extensible Mark-up Language

ICES, 35–6
ICMP, 74
IDA, 31
IEC, 248, 251
IIOP, 102
Image2000, 267, 269, 270, 271, 272
InfoSleuth, 41–2
Inland water application, 87–90
INSPIRE, 23, 25, 34, 37, 289, 290, 291
Interaction model, 109
IOW, 93, 97, 115, 126, 140, 147, 150, 152, 155, 161, 162, 181
IOW database, 120, 126
ISO/IEC 11179, 248, 251

JADE, 12, 69, 85, 110, 115, 118, 121, 143, 176, 180, 182, 185, 193, 194, 195, 225, 226, 230, 241
JDBC, 56, 57, 70, 71–2, 76, 79–80
JRC, 13, 218

Knowledge Browser (KB), 124
KOS, 248
KQML, 64, 65
Knowledge Recognition (KR), 252, 254

LDV, 89–90, 114–15, 120, 127, 138–42, 143
List of determinands, 147–55
Load/pressure indicator, 210–11
Local Database View (LDV), 89–90, 114, 120, 127, 139–41

Map interface, 198
Map services, 213–24
Mapping, 139–41
MapServer applications, 221, 223
Matchmaker agent, 43, 44
MCC, 39, 41, 42, 48, 83
MDS, 146, 162
Meta-* factor, 256
Molodenksi transformation, 270
Multi agent system (MAS), 63–6, 97, 100–107, 111–113
Multilingual List, 95, 171
Multilingual user interface, 199–201

NERI, 97, 115, 126, 129, 140, 149, 152, 155, 162, 225
NERI database, 129

OECD, 32
OGC, 25, 213, 214, 216, 221
OLM, 245, 252
Ontological model, 45, 123
Ontology agent, 43, 155
Ontology mapping, 139–42
Ontology provenance, 138
Ontology services, 118–20, 139–42
Orientation layer, 216, 218
OSPAR, 31, 35, 36
OWL, 6, 23, 25, 63, 90, 104, 128, 129, 130, 131, 132, 135, 139, 140, 143, 249, 254

Packet filter, 73–4
Performance time model, 239–41
Pfafstetter code, 274–5
Processing times, 227, 230, 232, 234, 236, 238, 239
Projecting in time, 207
Projection systems, 267, 269–72
Prototype, 84, 87, 89, 90, 95, 96, 115, 188, 193, 199, 204, 218, 221, 222, 225–6, 235, 241, 293

Query handling, 44

Radar sensing techniques, 263–5
RDBMS, 52–4, 70–3
RDF, 61–2, 63
Resource agent (RA), 43–5, 75–80, 113, 114, 117–21, 240
Representational State Transfer (REST), 26, 216, 222
Results interface, 201
River basins, 13–14
RMI, 76–8, 79
RNDE, 18
ROD, 30, 33, 35
Role model, 109–10

SANDRE, 19–21
Screened subnet, 74–5

Search interface, 198–9
Search times, 234, 240, 241
Security mechanism, 70–1
Semantic management, 94–5
Semantic model, 127–43
Semantic Web, 59–63, 254
Shannon index, 281, 282
SIE, 17, 18, 19, 24, 25, 26
SMTP, 25
Sniffer agent, 176
SOAP, 25, 26, 58, 181, 216, 222
Spheroid, 271
SQL, 52–4, 56, 141–2
SRTM, 260, 263, 264, 265, 267, 269, 273
SSL, 80
SSOA, 245–56
Statistical values, 206–207
Strahler order, 278, 281, 282

Task Agent (TA), 86, 113, 116–18
Task model, 108–109
Terminology Reference System (TRS), 6, 95, 145, 147, 150, 152, 155, 159–162, 169, 173
Time series analysis, 206–9
T-REKS, 138

UDP/IP, 64
UK-HMS database, 225
UML diagram, 91, 109, 176
UNEP, 146
Unit Ontology, 134–5
US-DOD, 45, 83
US-EPA, 6, 240
User agent (UA), 43, 86, 113, 116, 195, 230
User Interface, 186–7, 199–201, 246
User management, 198, 201–202
USGS, 264
UTM, 270, 271

Water Framework Directive (WFD), 13, 23, 29, 36, 94, 149, 162, 259, 261, 272, 287, 288, 292
Water reference, 259
Water surfaces, 266–72
WATERBASE, 97
Web application, 193–8
Web user interface, 193
WFS, 214, 215, 216, 222
WGS, 269, 271
Water Information System for Europe (WISE), 290–6

XMDR, 245, 249